Unforgettable
islands to escape to
before you die

Unforgettable
islands to escape to
before you die

Steve Davey
and Marc Schlossman

BBC BOOKS

CONTENTS

By their very nature islands are often unforgettable. Some were forged by volcanic eruptions, others thrust up by tremendous forces deep in the earth, cut off by floods or erosion, or torn from the nearest continent by the inexorable movement of the earth's tectonic plates. Their stark beauty and peaceful solitude bear witness to the drama of their creation. In their isolation, they may have developed species, ecologies and even weather patterns of their own. Their inhabitants, too, are often unique: cut off to some degree from cultures. Whether distant or almost part of the mainland, they will have evolved their own lifestyle and customs.

Whatever island you may choose to visit, travelling to it is easier than ever. You can now fly to almost any country in the world in a day; in two, you can reach all but the most remote destinations. Within a mere 72 hours of leaving home you can find yourself at the very ends of the earth. I delight in the fact that for less than the cost of a plasma TV set it is possible to take a week or two off work and go somewhere truly stunning. All it takes is the will. And it is in that spirit that we bring you this book. Instead of buying the TV set, you buy an airline ticket. Instead of going to work, you go to the airport. Instead of spending a week at your desk, you spend a week in paradise. Waking up to an early morning alarm, ready to head off to some far-flung destination, I often consider the simplicity of my options. I could switch off the alarm and stay in bed, or I can get up, make that familiar trip to the airport and be in the middle of nowhere tomorrow, eating noodle soup in Laos, drinking tea in Yemen or bathing with pilgrims in India.

Since *Unforgettable Places to See Before You Die* – my first book in this series – was published I have met or been contacted by a number of people who have been inspired by it. Some have given it to a partner as a subtle call to action. And at least one person, given it as a retirement present, has decided to try to visit every place in the book.

For this fourth title, my associate photographer Marc Schlossman and I travelled to all 40 of the islands featured in just 11 somewhat frenetic months. (The map on pages 252–3 shows their locations.) What I hope we have achieved is another snapshot of this beautiful planet that will encourage people to experience more of it, and to consider visiting some places they would otherwise never have considered.

At the start of the project I decided not to repeat any of the islands covered in the first three titles in the series. You won't find the Galapagos, Santorini, Manhattan, Zanzibar, the Maldives, Cuba, Iceland, Ireland, New Zealand's South Island, South Georgia, Aitutaki in the Cook Islands, or the islands of Australia's Great Barrier Reef in this book, but in a sense they should be considered an extension of it. Instead, we have chosen to include such eclectic, out-of-the-way destinations as Sagar in India, Lamu off the coast of Kenya, Socotra in the Arabian Sea and even Rapa Nui (Easter Island) in the Pacific Ocean. Other islands, such as Madeira, Ibiza, Sicily or Bali, might seem more familiar, but we have tried to show sides of them you might be unaware of.

About halfway through this year of photography I learnt that my partner and I were expecting our first child. I found out in Botswana during our visit to the salt-pan island of Kubu. Since then I have looked at the world with fresh eyes, and it has seemed more fragile than ever. I wonder whether my daughter will have the opportunity to gaze upon the sights I've seen; whether she will see dolphins and elephants, baobab trees and unspoilt beaches. Will vibrant cultures still exist, or will the homogenizing effect of globalization and proselytizing religions have crushed their individuality?

Some people think tourism endangers local cultures, but if it is done with careful consideration, it can both encourage and reward their preservation. While tourism can be a positive influence, it would be naïve to believe there are no negatives. Islands are very susceptible to mass tourism, and to one of its latest trends in particular: huge cruise ships that enable people to travel through the earth's less developed regions in a four-star comfort zone without ever coming into direct contact with their surroundings. Unlike the little ships that took us around Tierra del Fuego, and Svalbard in the high Arctic – two of our more far-flung destinations –

these liners churn sometimes thousands of passengers around the world in an air-conditioned bubble.

There is a lot you can do to make the world a slightly better place. Remember that your holiday destination is someone's home and that they would appreciate your respect, courtesy and a friendly greeting. Don't avoid big hotels, but ask them about their environmental policies – preferably before you reserve a room. Throughout the world, many admirable people and companies involved in tourism have been caring for their environment and local communities for years, others are just jumping on the save-the-planet bandwagon for marketing reasons.

Above all, travel softly and gently and engage with your surroundings. I have shaken so many hands over the past year, looked countless people in the eye and greeted them with a simple smile. I have been rewarded with unfailing hospitality, curiosity and generosity in some of the poorest, most distant places on earth.

Steve Davey, 2007

Palm trees on an idyllic Huahine coast

Ever since the first European explorers reached Tahiti, the Society Islands of French Polynesia have been synonymous with paradise on earth – a combination of gorgeous beaches, friendly locals and peaceful idyllic lifestyles. This impression was enhanced by the paintings of Paul Gauguin, who lived on Tahiti in the 1890s, and it still rings true today.

Colourful fish on the reef at Huahine

The 15 or so islands in the group are dominated by Tahiti. Most trade and industry happens here, and many people think it is a country in its own right. However, although it has many attractions – in particular, its rugged and mountainous interior – for a classic white beach and clear turquoise water speckled with deserted *motus* (coral atolls), you should head for other islands in the group. Iconic Bora Bora is among the best known, but therefore one of the most visited, so for a completely different experience try nearby Moorea and Huahine.

Just a short ferry ride from Tahiti, Moorea is a clutch of eight mountains that plunge into a lagoon of iridescent turquoise, fringed on one side by tall palm trees and on the other with breaking waves from the Pacific Ocean. On its placid waters you can sometimes see the crews of outrigger canoes training for races.

Lagoon at Moorea

Traditional outrigger canoe off Huahine

The lagoon is sprinkled with *motus* and it is worth taking a boat out to explore them – and feed the rays that will swim straight up to the beach when you arrive. They are so used to being fed that if you sit in the shallow water they will 'monster' you for attention. The *motus* are tropical paradises and leaving them at the end of a day of ultimate relaxation is difficult.

Huahine is also fringed by a turquoise lagoon. Smaller and much less developed than Moorea, it has a truly local feel. There are a number of bars and restaurants in the small settlement of Fare

Motu in the Moorea lagoon

Verdant mountains on Huahine

Tropical plant, Huahine

where the island's thriving community congregates after sunset. Food stalls are set up in the small car park and impromptu singing sometimes breaks out.

Stunning as Huahine is above the water, the undersea world is equally spectacular. The snorkelling in the lagoon is breathtaking and there are fantastic dives around the reefs at its edge. A local organization attracts black-tipped reef sharks by feeding them; they are relatively harmless, so you will be able to swim with these sleek yet powerful animals – a thrilling experience. Other fish also enjoy the free meal and dart around in great, brightly coloured shoals.

The many archaeological sites on Huahine give an idea of the complex ancestry of the people who live on the island. Most of the ruins ring the coast, but the largest and best preserved are the sacred *marae*, on the shores of Lake Fauna Nui, and at the foot of Mount Mouatapu. Maeva Marae was the seat of Huahine royalty and most important of over two hundred stone structures and *marae* of the island.

Predictably, as part of French Polynesia – formerly an 'overseas territory' of France and now designated an 'overseas country' – the Society Islands have a distinctly French feel and a French legal system; they are effectively part of France and depend on it for economic support. Their main revenue comes from the export of the black pearls that are farmed throughout the islands – beautiful souvenirs of a visit to a tropical paradise.

(i) ··

Air Tahiti Nui flies to Tahiti from several European cities, including London and Paris. Air Tahiti (the Tahitian domestic airline) offers a range of internal flights, and Moorea is a short ferry ride from Tahiti. Huahine can be reached via a short domestic flight. The cost of living is high in French Polynesia, and it is worth looking for a package to secure a kinder hotel rate. LAN Chile flies from Tahiti to Rapa Nui (see page 246) twice a week, which allows you to combine two unique destinations.

Looking across to Moorea from the island of Tahiti

Lagoon at Huahine

Si Phan Don

Laos

In the sleepy deep south of Laos, just before the border with Cambodia, the Mekong River pauses in its headlong dash to the South China Sea to sprawl across a 14-km span, creating hundreds of tiny islands. This inland delta is optimistically called Si Phan Don (Four Thousand Islands).

Some islands are just rocks poking out of the water, but others, such as the main one, Don Khong, are more developed with villages, roads and, of course, the ubiquitous wats – Buddhist monasteries or temples.

Wat overlooking Muang Khong on the island of Don Khong

Fishermen in the Mekong off Don Khong

Monk on his morning alms round at sunrise

Laos is a Buddhist country and every morning, in the most atmospheric and endearing ritual in South-east Asia, saffron-robed monks from local monasteries head out on their alms round. In the sleepy provincial capital of Muang Khong on Don Khong, they can be seen walking through the streets collecting donations of food in their characteristic begging bowls.

South of Don Khong, close to the islands of Don Deth and Don Khone, lie two waterfalls, part of a series of rapids that block the Mekong and brought to an end the French colonial dream of

constructing a navigable route through the countries of Indo-China. To circumvent these, engineers built a railway that stretched across the islands. There was a loading gantry at either end of the line, and a passable copy of the stone bridge at Avignon still spans the waters between Don Deth and Don Khone. Such a solid structure looks incongruous in this rural location.

Traditional houses overhanging the river at Don Khone

Rusting locomotive

The railway has gone, but from Don Deth you can walk along the old line and across the bridge to Don Khone, where you will find a decrepit and rusting locomotive stranded on a few lengths of gnarled rail.

The Khong Papheng Falls are the largest and most spectacular of the two waterfalls, and the best time to see them is in the early morning or late afternoon when fishermen climb over boiling cataracts on rickety, log bridges to cast their fishing nets. One slip

Sunset over the channel between Don Khone and Don Deth

Fishing by the Khong Papheng waterfall

Monks on their morning alms round

and they would be swept into the water, but in the languid way that makes Laos so compelling they never seem to rebuild these makeshift structures.

People cast fishing nets all over Si Phan Don, with dozens of them fishing from the small wooden canoes that congregate as the heat of the afternoon slips away. The best way to appreciate the sight is sitting in one of the handful of tourist cafés that overhang the river in Muang Khong, while sipping a legendary Beer Lao and watching the far bank of the Mekong River glow golden in the sunset.

Try to split your time between Don Khong and Don Khone. The former is more sophisticated, while Don Khone has a quiet rural charm, with a few colonial-era buildings dotting its shady dirt streets. Avoid Don Deth, though, which has recently been blighted by a rash of shoddy, identical backpacker huts.

At the far end of Don Khone, about a 30-minute walk along the old railway line, is Ban Hang Khone, home of the French railway's southern loading gantry and the base for excursions to see the rare freshwater

Irrawaddy dolphin. Just turn up at the village in the mid to late afternoon and ask for a boat. Its owner will ferry you out to a small island in the Mekong where you can look out for these shy and elusive creatures.

The dolphins are protected, but their numbers are dangerously low and it takes patience to spot one – usually a long way off. The most you can hope to see is a spindly dorsal fin breaching the water a few hundred metres away. The best time to look for them is in the late afternoon, when they often appear in front of the local boats that weave around on the river. The looming hills of Cambodia, across the Mekong, are an atmospheric backdrop in the setting sun.

With the opening of the Cambodia–Laos border to tourists, getting to Si Phan Don has never been easier. You can travel from Phnom Penh in Cambodia or from the Lao capital, Vientiane, via Pakse. If you are approaching from Thailand, Etihad fly from London to Bangkok. You can then take an internal flight to Ubon Ratchathani and cross the border to Pakse. Audley Travel has had many years of experience in the region, and can tailormake an itinerary that follows any of these routes. Having a guide and a driver will help you to avoid the languid Lao public transport. At weekends day trippers from Thailand can spoil the calm.

Woman selling snacks – including cooked frogs

Sunset from Don Khong Island

Newfoundland
Canada

Just off the east coast of Canada, Newfoundland is known as the 'crossroads of the world' to its inhabitants – and it lives up to its nickname. As the island is closer to Europe than any other part of the American continent it was – and is – strategically important for travel, navigation and communications.

Lobster Cove Head Lighthouse

The Vikings established the first non-indigenous settlement in North America here, and Amelia Earhart took off from Harbour Grace when she became the first woman to fly the Atlantic in 1932. The first trans-Atlantic cable was laid to the small village of Heart's Content in 1866; and Marconi received the first trans-Atlantic wireless message at the aptly named Signal Hill in St John's, the island's capital, in 1901.

Today, if you are flying between Europe and North America the chances are that the striking scenery you see out of the plane window when you're close to Canada is Newfoundland: the curvature of the earth's surface means the shortest flight path between London and New York often takes you over it. The *Titanic* was following a similar route when she struck an iceberg less than 640 km from the island –

Bonne Bay, town of Woody Point and Tablelands, Gros Morne National Park (background)

Cape St Mary's Ecological Reserve

its Cape Race Lighthouse was the first to receive the SOS signal from the stricken ship.

Because of global warming there are fewer icebergs than there were in the days of the *Titanic*, but they can still be seen in June and early July. The convergence of the northbound, warm Gulf Stream and southbound, colder Labrador Current off Newfoundland drags in icebergs that have calved from the Greenland ice shelf. The two currents also bring in vast amounts of plankton as well as fish and the whales that feed on them – humpback, fin and minke whales are often seen in the island's offshore waters.

On land, Newfoundland is home to the highest concentration of moose in North America, and there are caribou on the Avalon Peninsula. The island's birdlife is spectacular, especially in the Cape St Mary's Ecological Reserve, which has some of the most accessible seabird colonies in the world. In the breeding season, from April to

Woody Point Lighthouse

June, vast numbers of northern gannets, kittiwakes and common murre form noisy garrulous colonies. Rarer species that nest here include black guillemots, razorbills and double-crested cormorants. It is possible to get to within a few metres of the birds.

The most dramatic region of the island is arguably Gros Morne National Park. Its landscape is the result of a combination of glaciation and tectonic plate movements, and the Western Brook Pond is set in a classic example of a valley that was carved out by a glacier. A boardwalk leads through a series of bogs (where you will find orchids, carnivorous plants and even moose) to the so-called pond – it is actually 3 km long. A boat tour will take you to where it narrows through a steep gorge with sheer cliffs and peaks that rise to 650 metres. The Tablelands area was formed by tectonic movements that forced peridotite up from the ocean floor. This part of

Tablelands, Gros Morne National Park

Fair and False Bay, Burnside

The Battery, St John's

St John's

the park is barren, as peridotite, rare at the earth's surface, inhibits the growth of vegetation, yet nearby slopes are forested.

There are a number of remote settlements on the Newfoundland coast. Known to the local people as 'outports', some are small clusters of weathered dwellings but others are complete villages, and visiting them is like stepping back 100 years in time. These are not just tourist traps; they are well-preserved working fishing communities, even though the cod industry has been hard hit by dwindling stocks. Trinity on the Bonavista Peninsula is a particularly atmospheric outport.

The oldest city in North America, St John's is lively and colourful, with a vibrant feel that contrasts well with the island's unspoilt natural scenery and long cold winters. It has numerous bars and nightclubs – George Street, with 28 on each side, has the highest concentration in the northern part of the American continent – and the houses that line many of the city's other streets are painted in different vivid colours.

Hatchet Harbour

 ··

Only Air Canada flies to Newfoundland so flights are relatively expensive. You need to hire a car well in advance, especially if you visit the island during the peak summer months. Be aware that it is not possible to get unlimited mileage on your car hire. Distances are vast and extra mileage charges can mount up.

The Île de la Cité may blend into the Paris rooftops when viewed from the top of the Eiffel Tower, but the island is one of the most significant places in France. Politically, legally and religiously it lies at the centre of the country and, by extension, the centre of the French Empire that once controlled a number of the islands in this book.

Most visitors to Paris never think of the Île de la Cité as an island. There are certainly few clues to this as you stroll over the stylishly expansive Pont Neuf, one of four bridges that span the Seine on either side of it – and none if you arrive at the Cité metro station.

Île de la Cité and Pont Neuf from the Pont des Arts

Most historians believe the island was first settled in 52 BC by a small Celtic tribe, the Parisii, and it is from this original inhabitation that the city of Paris developed. Today the Île de la Cité is the site of one of the most famous cathedrals in the world: Notre Dame. The menacing Conciergerie, the infamous prison of the French Revolution is also here, as is the Palais de Justice and, opposite it, the shops selling barristers' gowns.

View from the balcony of Notre Dame cathedral

Notre Dame has to be the best known of these buildings, and certainly draws the most tourists. Day and night they pose in front of the cathedral's instantly recognizable façade with its twin bell towers

and gargoyles – made all the more famous by Victor Hugo's *The Hunchback of Notre Dame*. It is worth joining the long queue to climb to the top of one of the towers and see the bell Quasimodo rings in the novel. Panoramic views of Paris stretch away on all sides.

The interior seems larger than it really is, and can absorb any

Street scene behind Notre Dame

number of visitors and awe them into silence. The space is ringed by small chapels and ornate wooden choir stalls fill much of the east end.

Behind Notre Dame cathedral is the largely subterranean Mémorial des Martyrs de la Déportation. This simple yet harrowing structure commemorates the 200,000 people who were deported from Paris and murdered in the German death camps during the Second World War.

For all the grandeur and officialdom centred on the Île de la Cité, the island has a remarkably human face. The Vert Galant at the west end is a small and pretty garden, crowned by a quay, that is popular with lovers and picnickers alike. And the Place Dauphine is one of the simplest yet most quintessentially French squares in the city.

Looking over to the Eiffel Tower, with the tip of the Île de la Cité in the foreground

Buildings on the Île de la Cité

Triangular because of the way the island tapers, it is incongruously faced on one side by the white edifice of the Palais de Justice. Boule players duel amid stuntedly shady trees and a couple of small restaurants serve alfresco food in summer.

There is also a large flower market, which is replaced by a bird market on Sunday mornings, when the air is filled with the trilling of songbirds. Thousands of these are displayed in small wooden cages

ÎLE DE LA CITÉ

Tomb inside Notre Dame

Rose window, Notre Dame

Notre Dame and the Île de la Cité at night

Bookstall on the bank of the Seine with Notre Dame behind

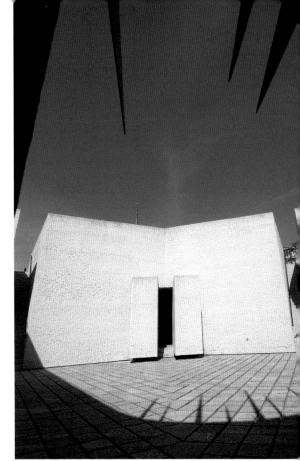

Mémorial des Martyrs de la Déportation

in a scene that is more reminiscent of Asia than the centre of a European city.

Connected to the Île de la Cité is the smaller and far more human-scale Île Saint-Louis. It boasts exclusive residential properties and its main street, with its bars and restaurants, still has specialist shops selling cheeses, antiques or ice cream, as well as ones that deal in the more usual tourist souvenirs.

Both islands come alive at night when every quay is filled with tourists and locals just hanging out; but they are at their most atmospheric in the misty early morning, before traffic noise and pedestrians take over a scene as timeless as the softly running Seine itself.

Eurostar will take you from central London to central Paris, currently in just over 2½ hours. A number of low-cost airlines offer flights from other UK cities, but check that they fly to Charles de Gaulle airport, not distant Orly.

Islands of Lake Tana
Ethiopia

Misty morning on the shores of Lake Tana

Papyrus boat fisherman being followed by pelicans

More than half of Lake Tana's 37 islands are home to a monastery or church, and visiting them is an enthralling way to appreciate Ethiopia's rich Christian heritage.

Although the country is, for many people, synonymous with famine and drought, it arguably has more history and historical sights than anywhere else in sub-Saharan Africa. It is Christian, but unlike many other African countries the religion was not foisted on it by European missionaries. It adopted Christianity in the 4th century AD and developed its own Coptic Church, with an art and history that are relevant to its people. Among the most touching examples of this are ancient murals that show Mary suckling the baby Jesus – because, as

anyone who spends even a short time in Africa will realize, that's what mothers do!

Taking a boat across Lake Tana's muddy brown waters, from Bahar Dar on its southern shore to the island of Tana Cherkos on its far side, is like stepping back in time. As the sun rises through the morning mist, fishermen paddle out in traditional *tankwa* (papyrus reed boats) to check their nets. The same craft can be seen in ancient

Monk on the cliffs of Tana Cherkos

Portuguese Bridge on the mainland near Bahar Dar

Egyptian wall paintings – Lake Tana, in Ethiopia's north-west Highlands, is the source of the Blue Nile and the country had early links with Egypt. The boats are often followed by pelicans who try to snatch the fish and wait for the odd scrap to be thrown to them.

Tana Cherkos is probably the most historic of the lake's islands. Ethiopian lore says the son of King Solomon and the Queen of Sheba (who Ethiopians believe came from Axum in the north) brought the Ark of the Covenant back from the Holy Land. It was apparently hidden on Tana Cherkos for 800 years before being moved to Axum, where it is reputed to remain to this day – meanwhile, it seems, the distrust of strangers engendered in its guardians survives in the monks who still live on the island. As we arrive, a number of them stare at us from its high cliffs. The welcome isn't much more friendly once we land. The monks show us the historic painted church and their store of treasures, including an ancient pre-Christian dish for animal sacrifices, although I get the distinct feeling they are keeping the real stuff hidden.

Cataract of Tisissat (Blue Nile) Falls

Shoreline of Tana Cherkos

Priest leaning by the solid wooden door of Narga Sellase

Ancient painted interior of Tana Cherkcs

All the island churches and monasteries have collections of historical artefacts, most of which can be seen for a small contribution. The treasures range from Ethiopian crosses and the crowns of the country's various kings, to illuminated bibles, handwritten in the ancient religious language of Ge'ez. At the 11th-century Dega Estanfanos monastery you can even see the mummies of kings and emperors, including that of the 13th-century emperor Yekuno Amlak. There is no electricity on the island, and you are ushered into a small and stifling room by a priest who shows you the relics by candlelight.

Although many of the monasteries are closed to women a number are open to everyone. These include Ura Kidane Mehret on the Zege peninsula, Narga Sellase on the island of Dek and the Intoes Iyesus convent. The Ura Kidane Mehret monastery is one of the most ornately decorated on Lake Tana, with paintings showing scenes from Ethiopian history and religious tales. Although they look as though they were created yesterday, they are in fact hundreds of years old, but the natural dyes and pigments the artists used have resisted fading. Although the inner sanctum looks quite plain from the outside, it is completely covered with murals. The work of generations, they are made all the more poignant by their remote and rustic setting.

There is a similar riot of colour on the walls of the inner sanctum of the Narga Sellase monastery where, for a small donation, its somewhat roguish monks will break out the contents of their small museum: crosses, crowns and a 17th-century bible.

It takes at least two or three days to see all the monasteries and churches on Lake Tana, as the distances can be long and the boats are atmospherically slow. If you feel like a break, the Tisissat (Blue Nile) Falls are about an hour's drive from Bahar Dar. Sadly emasculated by a hydroelectric plant, they are still worth a visit, if only to see the 17th-century Portuguese stone bridge over the lower reaches of the river.

ⓘ ···

Ethiopian Airways flies to Addis Ababa. It has a full domestic network that will take you to many of Ethiopia's historical cities, including Bahar Dar. Organizing land arrangements can be difficult – especially if you are short of time. Journeys By Design are specialists in African travel. As the name suggests, they will create custom itineraries that take in Lake Tana.

Priest showing off ancient treasures

Priest displaying an Ethiopian cross at Narga Sellase on Dek Island

Rocky scenery from the Balcões near Ribeiro Frio

Terraces in the interior of Madeira

Long synonymous with sweet wine and winter breaks for the elderly, Madeira is also a paradise for hikers and nature lovers. The tip of an underwater mountain range, it shoots abruptly out of the Atlantic Ocean some 950 km from the Portuguese coast.

The island is only 57 km long and 22 km wide but its highest point, the craggy and atmospheric Pico Ruivo, rises 1861 metres above sea level. This sudden height amid the wet waters of the Atlantic creates microclimates. Although the temperature hovers in the 20s all year round, Madeira's mountainous centre is often enveloped in rolling impenetrable cloud that, seemingly magically, gives way to hot sunshine before the mist sweeps in again. And the weather can be different on each quarter of the island: if it is raining in one area drive over a pass, or through a road tunnel, and there is a good chance you will be bathed in sunshine.

Rugged coastline on the north of the island

Peaks at Pico do Arieiro

Coastline at the foot of Reid's Palace

The starting point for treks to Pico Ruivo is the atmospheric peak of Pico do Arieiro, reached via a switchback road with many hairpin bends. From here you look across seemingly endless knife-like jagged peaks and out to the sea nearly 2 km below. Between the rock-blades deep valleys slice into the island – their floors see daylight only at midday when the sun is high enough to pierce their hidden depths. The walk is only 4 miles, but includes a number of vertiginous climbs and descents of over 450 metres.

Not all the hikes are as strenuous as the trek to Pico Ruivo. Many trails follow Madeira's *levadas* (watercourses), which date back to 1452. Cut into the contours of hills, they were developed to move water around the island and sometimes run for dozens of kilometres. They have a gradient of only a few degrees, so the water always flows – but not too fast. Alongside the *levadas* are walking tracks, some a few metres wide, others just narrow ledges cut into a hillside. There are also long tunnels hacked through rock, initially by slaves brought to work on sugar-cane plantations.

Steeply terraced village of Curral das Freiras

Jagged mountains at Pico do Arieiro

Plants on the summit of Pico do Arieiro

As well as being the island's main thoroughfares as recently as the mid-20th century, the *levadas* allowed seemingly inaccessible places to be irrigated for agriculture. The result can be seen in a key feature of the Madeira landscape: the *poios* (terraces) on many of the mountainsides. Nowhere on the island is there a better example of this terracing than in the remote Curral das Freiras (Nun's Valley). Viewed from the top of the surrounding hills, seemingly sheer walls drop to the gloomy valley floor below. There are *poios* most of the way down and the effort required to bring these relatively small areas under cultivation must have been phenomenal. The sun reaches the village at base of the valley for only a few hours a day, and it is a shame that *levadas* can't bring in sunlight as well as water.

Funchal is Madeira's main town and it is here that you will find Reid's Palace, one of the world's classic hotels. Perched on cliffs overlooking the sea, it has quiet gardens, tennis courts and three heated saltwater pools. Once patronized by Winston Churchill, George Bernard Shaw and a host of kings and queens, this is old-world luxury. Even if you can't afford to stay here you should sample afternoon tea on the terrace overlooking the sea.

Funchal is famous for the wicker toboggans that run down from Monte in the hills above it. Not quite a white-knuckle ride, they clatter noisily through streets polished over the years by their runners, propelled by two *carreiros* wearing straw boaters. A cable-car ride to the top of the hill will give you commanding views of the town and its surroundings.

ⓘ ··

British Airways fly direct from Gatwick to Funchal. During the summer season a charter flight is the best option. Hiring a car is vital if you want to explore. Distances aren't great so it is possible to base yourself in one place. The doyen of Madeira hotels is the historic Reid's Palace. Owned by Orient Express, it occupies the best viewpoint in Funchal.

Cabo Girão, the second highest sea cliffs in the world

Looking out to Ilhéu de Farol from Pico do Arieiro

Sagar
India

India is intense. Eternal. Nowhere else in the world is there the same balance between ancient and modern. On the one side are industrialists and computer programmers; on the other a spirituality that permeates and orders the very nature of society. And nowhere is this more apparent than on the island of Sagar during the Ganga Sagar *mela* (festival) at Makara Sankranti.

This is one of the great gatherings that are a key feature of Hinduism, India's main religion, during which pilgrims bathe in sacred waters to wash away their mortal sins and so free themselves from the endless cycle of reincarnation. Sagar is in West Bengal, near Kolkata

Ferries sailing to Sagar island for the *mela*

(Calcutta), at the point where the sacred River Ganga (Ganges) reaches the Bay of Bengal. And in mid-January each year over half a million pilgrims travel there to celebrate Makara Sankranti at the start of the sun's journey to the northern hemisphere.

The ethereal nature of the island is accentuated by a night-time fog that often lasts long into the morning. Swirling and all-enveloping, it is coloured orange by sodium lights, and green, blue and red by coloured ones that demarcate the main roads so that the many – often illiterate – pilgrims can orientate themselves.

Although people bathe a couple of days before Makara Sankranti, the most auspicious time is on 14 January, just after midnight. Streams of pilgrims make their way to a beach where the Ganga is said to meet the sea. The real confluence is slightly further away, but currents

Pilgrims making offerings

Pilgrims bathing in the sea

Pilgrims at the Ganga Sagar *mela*

make entering the water dangerous and during the *mela* the authorities forbid bathing there and ground the local boats to enforce the ban.

It is impossible not to be affected by the sheer number of people, many of whom have travelled great distances, and all of whom share a single belief and intent. The spirituality is palpable and the feeling of euphoria infectious. Hinduism is a joyful religion and there is little solemnity at the bathing area. People smile and splash each other. They dare each other to duck under the waves for a full immersion and surface, spluttering and laughing. Smiling, happy, cleansed.

Walking through this throng you will not feel excluded or unwelcome. The pilgrims draw you in far more than celebrants of any other religion would. And the pressure to bathe is irresistible – people find it inconceivable that you would come all the way to Sagar and not do so.

After they have immersed themselves in the cold water the shivering pilgrims make their way up to the Kapil Muni Temple, where they crowd around the priests, attempting to give them offerings in return for *prasad* (sugary sweets that have been blessed).

Pilgrim praying in the sea

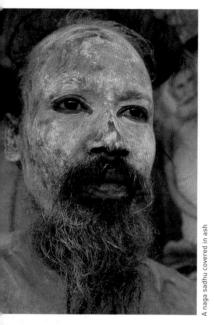

A naga sadhu covered in ash

As they move away from the temple they walk along a road lined with the cell-like ashrams of naga sadhus. These ascetic holy men, who are often considered to be living saints, eschew material possessions including clothes – all that covers them is ash from fires. In prudish India the fact that they choose to walk round naked is taken as being completely normal. They also sport long dreadlocks and smoke *charras* (hashish) when they meditate.

Naga sadhus are renowned for their extreme approach to spirituality: yoga, fasting, days of meditation. Some even perform standing penances and remain on their feet for years at a time. Although they are primarily at the *mela* for religious reasons, Ganga Sagar is also a peak time for business. They bless the pilgrims in return for a few rupees – making the money that will see them through until there is another festival to attend. Sagar receives pilgrims throughout the year, and has a semi-permanent contingent of holy men, but during the *mela* their numbers are swollen in that uniquely Indian, indistinguishable blend of the spiritual and the commercial.

Kapil Muni Temple

Crowds of pilgrims on the temporary main street of the *mela*

Pilgrims pushing to make offerings at the Kapil Muni Temple

Pilgrims struggling to receive *prasad*

Indian festivals are like this. Amid all the spirituality there are sprawling bazaars and even fairground attractions. Many pilgrims invest what to them is a phenomenal amount of money to get to Sagar, and they combine their visit with shopping for goods and clothes that are not available at home. As the *mela* winds down they board buses that carry them to the embarkation point and the ferry boats that will take them to the mainland. Some of the more devout pilgrims (and the poorer ones) walk 30 km or so across the island.

At times the site of the festival resembles a great refugee camp. People plod great distances to reach it and many sleep rough in the cold January nights of West Bengal. They are sent on long diversions to avoid potentially fatal crushes and endure many hardships in addition to their immersion in the bitterly cold waters of the Bay of Bengal. But Ganga Sagar is a happy and fulfilling time. A time that will reward the faithful and the curious alike with memories of one of the world's last great pilgrimages.

Ferry arriving at Sagar island at sunset

(i) ....

The *mela* on Sagar takes place on 14 January every year. Jet Airways flies to India from many international airports, including Heathrow, and connects with their extensive domestic network, which has daily flights to Kolkata. From here the easiest way to get to Sagar is by private car. Accommodation at the *mela* is very basic and should be booked well in advance. Atithi Voyages in Delhi can make all the arrangements.

Forested coastline of Male Kula

The landscapes in the Vanuatu archipelago in the south-western Pacific are dramatic: brooding rainforests cover many of the 83 islands, some coasts are fringed with black sand beaches, and active volcanoes, such as the frenetic Yasur on Tanna, malevolently spit out lava.

Island off the coast of Male Kula

But Vanuatu is a remarkably friendly place. Although the sight of groups of men carrying half-metre-long machetes and taking exploratory thwacks at random palm trees may take some getting used to, they will probably greet you with a cheery 'Good morning' or 'Bonjour'. The country was jointly colonized by Britain and France, and a family's choice of school for its children depends largely on its Anglo- or Francophile leanings.

Palm trees on the island of Ambae

On the surface this island nation (formerly the New Hebrides) resembles many other remote places that have undergone the combined onslaughts of colonialism and missionary zeal: fragile cultures lost in the curious belief that to be a Christian it is necessary to dress and act like a European. But in Vanuatu the old customs – *kastom* in the local Bislama language (a spoken pidgin English with an 8000-word vocabulary) – seem to be always present in the lives of even the most devout *ni-Vanuatu*. Growing tourism may have reinforced this, as traditional ceremonies can be a source of income.

Many villages on outlying islands have some sort of *kastom* celebration, especially around harvest time, but Vanuatu's most famous one is *naghol* (land diving), on the island of Pentecost. This

Nagbol (land diving) on Pentecost

Volcanic beach on Male Kula

Naghol (land diving) on Pentecost

Kastem village on Tanna

forerunner of bungee jumping involves the construction of a large tower that ranges in height from 15 to 30 metres, with platforms all the way up from which young men and boys throw themselves, attached to the tower only by a yam vine around each ankle. The vines are chosen for their strength and according to the weight of the jumper. Accompanied by chanting and dancing, the participants climb to platforms allotted to them according to their age and status. As each one reaches his level, the chanting builds to give him encouragement. Jumpers often posture before launching themselves into the air, throwing handfuls of grass contemptuously to the ground.

The towers are built on a hill, and as the jumper leaps, head first, the platform gives way, swinging him back into freshly dug and softened earth on the hillside and breaking his fall. They still hit the ground with a thump, though, and the crack of the platform as it collapses – although integral to the safety of the jump – is a haunting and disturbing sound. The *naghol* can be performed only in April and May, when the vines are still wet and supple after the rainy season. Famously, the *naghol* was once held outside of this season for the Queen, resulting in a fatality when both vines snapped.

Some places in Vanuatu have rejected western ways completely. I was lucky enough to visit Yakel village on the island of Tanna which has decided to follow its old customs. This was a conscious decision based upon pride in its local culture. The villagers are not some

remote tribe waiting to be snaffled up by the encroaching 'progress' of development – they are maybe an hour away from a local store that stocks Australian Cabernet Sauvignon for the tourist market. They have simply decided that, for them, their traditional ways are better. There are many similar villages, spread throughout the islands, but it is rare for them to be accessible – and for their inhabitants to welcome visitors.

Vanuatu is not the easiest place to travel around. There is a network of internal flights, but these are sporadic and often land on grass airstrips. On the outlying islands land transport is difficult and expensive, and accommodation is basic to say the least. Most visitors divide their time between the main island, Efate, and Espiritu Santo, which have more tourist development, and many come just for the scuba-diving they offer, which is among the best in the Pacific. But to ignore the culture of the islands is to miss out on a unique way of life that has managed to survive into the 21st century.

Storm cloud over an island at sunset

Outrigger canoe landing at Pentecost

(i) Air Vanuatu flies from Brisbane, Sydney and Auckland to the capital, Port Vila, on Efate. It also flies to Nadi in Fiji, which means you can fly with them to the Yasawa Islands (see page 114). Vanuatu-based Destination Pacific can make local arrangements for you.

Sunrise on Ambae

Clouds building over the peaks of the Red Cuillin

The Isle of Skye is the largest of the islands in the Inner Hebrides and is dominated by the mighty Cuillin Hills, which can be seen from all but its most remote parts. Many of their peaks are accessible without technical skills and they are justifiably popular with mountaineers and climbers.

The Cuillin are split into two groups. The Red Cuillin are composed of granite that takes on a characteristic reddish tinge at sunrise and sunset, and are round in profile. The Black Cuillin are made from basalt and gabbro, a rough, dark rock that gives them their name, and are far more striking. While they are relatively low – none of them reaches over 1000 metres – they loom impressively over the island. Craggy peaks and sheer cliffs form deep gullies and ravines, unsoftened by vegetation and often wreathed in cloud. The highest point on Skye is the summit of Sgurr Alasdair in the Black Cuillin.

The Cuillin are not the island's only spectacular feature. Much of the coastline is striking, with great cliffs plunging down to the sea or rolling farmland sweeping down to pebble beaches. On the most

western point of Skye is the Neist Point Lighthouse; painted yellow and white, an incongruous colour scheme in this remote location, it stands on a vertical cliff, the top of which is coated with meadows and wildflowers. Seabirds soar in the surrounding air currents.

The Storr, a rocky hill on the Trotternish Peninsula, is noted for the pinnacles at the foot of its steep cliff face, in an area called the Sanctuary. These are volcanic plugs, strangely shaped by the violence of their creation and erosion. The best known is the Old Man of Storr.

The wildlife on Skye is as spectacular as the scenery. The rare white-tailed sea eagle is sometimes seen here, and golden eagles are relatively common in the Cuillin Hills. In summer you may come

Western coastline of Trotternish seen from Sconser

Sound of Raasay as seen from Trotternish

Farmhouse at Drinan, Strathaird

Cemetery on the shore of Portree Loch, Portree

Red clover and lichens on the coast at Ord, Sleat

Pier on Armadale Bay at Isleornsay, Sleat

across dolphins, minke whales and even basking sharks in the island's lochs and inlets, and otters can be found all year round. On land there is a good chance of seeing majestic red deer.

Skye is steeped in history. There are cairns and standing-stone circles from Neolithic times – even dinosaur footprints at Staffin Bay – and it was known to the Romans and recorded by Ptolemy, the 2nd-century Greek geographer. The name itself is said to come from Old Norse, hinting at contact with the Vikings.

More recently, Dunvegan Castle has been the seat of the Clan MacLeod since 1237. It has been augmented by a keep and tower over the centuries, although most of the battlements were added in the 19th century. For all its fortifications – until the 18th century the only entrance was via a sea gate – the castle has seen little action, although it was besieged by the Macdonalds, the sworn enemies of the MacLeods, in the 15th century.

Although Skye is on the same latitude as Moscow, the Gulf Stream makes it much milder. However, the weather is changeable and you can be engulfed in a sudden downpour at any time, only to have the sun break through grey clouds a short while later. If it does rain you could visit the Talisker distillery at Carbost. Named after Talisker Bay, or possibly the Talisker River, it is the only single-malt distillery on the island. It dates back to 1840, and Robert Louis Stevenson mentioned it in his poem *The Scotsman's Return from Abroad*.

Gaelic and the culture it represents are highly valued on Skye – almost half the population speak the language and the island is sometimes called by its Gaelic name: Eilean á Cheò (Isle of Mist). There is also a strong heritage of folk music, and the annual Isle of Skye Music Festival in June attracts many well-known folk singers.

ⓘ ....

The island is easily reached by car or bus from Inverness and Glasgow over the Skye Bridge. There are train services to Mallaig and the Kyle of Lochalsh on the mainland side, and ferries to Skye and many of the nearby islands from Mallaig. Inverness and Glasgow have good rail and air links with the rest of the UK.

Kilmore, Sleat

Farm at Tormore on the Sound of Sleat, Sleat

Cousine and Praslin
The Seychelles

Clear waters break on the beach at Cousine Island

Long known as a tourist paradise, the Seychelles is also a haven for many endangered and endemic species. And nowhere more so than on Cousine Island, where all the profits from tourism are ploughed back into conservation and restoring the island to its original state.

The Seychelles is typified by massive granite rocks on small, golden sand beaches, and Cousine is no exception. The beach here stretches along half the island, and as there are only four villas, with a maximum of eight people at any one time, you should be able to find a spot for yourself.

Cousine Island

Except, that is, for the many uninvited guests: if you are here between September and December you will find hawksbill turtles lumbering up the beach to lay their eggs. Hatching takes place after two months, so if you visit Cousine in November or December you should be able to watch the eggs being laid as well as having a good chance of seeing a nest hatch. There is a full-time naturalist who is researching the turtles, as well as supervising the restoration of the island, and the nests are monitored and moved to a safer spot if necessary.

There can be so many turtles lurching out of the sea that they almost become an intrusion, forcing you to retreat down the beach so

as not to disturb them. Once they have started laying, nothing will bother them, and you can approach and watch their clumsy attempts to bury their eggs before they trundle back to the shore.

The hatchlings are a tremendous contrast to their parents in terms of size and could comfortably fit on their parents' head if they were ever to meet. Only one out of a hundred will survive to adulthood, yet they seem invulnerable as they swarm down to the water, clambering over anything in their path.

There are rare and endemic bird species all over Cousine. It is a significant breeding site for the Seychelles magpie robin and the Seychelles blue pigeon, and is home to over a thousand pairs of sooty terns and the world's largest colony of wedge-tailed shearwaters. Many of the birds are ground nesters, which is why all non-native animals on the island, such as rats, have been eliminated.

Giant spider on Cousine Island

Lizard on Cousine Island

Crab (above) and hawksbill turtle (below) on the beach at Cousine Island

Trees on Cousine Island

Cousine also has a number of giant tortoises. They are totally habituated to humans and will even siesta in the middle of a path – heads outstretched on the ground. Some are so tame that they will follow you, lumbering breathlessly in pursuit. There is no malice here, though, and their curiosity of human beings has long since expired. They just like their scaly, yet remarkably soft, skin to be tickled. Numbers have been painted on the tortoises' shells, so you can check their age, species and history against a list.

On nearby Praslin Island you can visit the Vallée de Mai – a nature

Praslin seen from Cousine Island

Cousine Island seen from Praslin Island at sunset

Giant tortoise on Cousine Island

reserve that is home to the coco de mer, an endemic species of coconut. The emblem of the Seychelles, it is the biggest coconut in the world and the local people delight in its resemblance to the female anatomy. It is even featured on the visa stamp in passports, where its imprint is similar to the one a rather large female nudist might leave in the sand. The reserve also shelters a number of other endemic plants and is the last patch of native forest on the island.

It is worth taking the time to explore Praslin and its beaches. Staying at the Lemuria Resort on the western tip of the island will give you access to two stunning private ones, and also to Anse Georgette, arguably the most beautiful beach in the Seychelles.

ⓘ ....

Air Seychelles flies to the capital, Victoria, on Mahe Island. Domestic flights are available from Victoria to Praslin, but Cousine can only be reached by helicopter transfer. Arrangements can be made through Turquoise Holidays.

Santa Barbara Islands
California, USA

Catalina Island

Santa Cruz Island

The Santa Barbara Islands (also known as the Channel Islands) lie in a scattered archipelago off the coast of California on the west coast of the United States. Although they are no further than 40 km from the mainland they have a unique ecosystem and are home to many endemic species.

During the last ice age, when the sea level was lower than it is today, many animals and plants moved on to land that became the Santa Barbara Islands. As the water level rose they were isolated and species evolved independently. In prehistory these would have included the pygmy mammoth, but now the island fox and pygmy jay are among the 145 unique animals that can be found here.

There is a rich ecological diversity under the sea too. The cold waters of the north Pacific meet a warm current from the tropics, which causes an upwelling of nutrients that supports a long food

Scorpion Ranch, Santa Cruz Island

Gull on Santa Cruz Island

chain, including vast kelp forests. The diverse habitats created by the continental shelf are home to many marine animals – sea elephants and sea lions are common and – over a year – 14 species of whale, including the California grey whale, can be seen around the islands.

Santa Barbara consists of eight main islands, five of which – Anacapa, Santa Barbara, Santa Cruz, Santa Rosa and San Miguel – are incorporated in the Channel Islands National Park. Each has its own unique character.

Anacapa has large bird colonies, including the country's largest rookery of brown pelicans, a species that was almost rendered extinct by DDT in the 1960s. Santa Cruz, the largest of the islands, is ringed by steep cliffs and the island jay can only be found here. There is also a programme to reintroduce the bald eagle to the island, which recently resulted in the first hatching in over 50 years.

Santa Rosa is notable for the Torrey pine, which only grows there, and the island oak. Its rolling hills are home to the island fox. San Miguel is the outermost island and the most exposed. The weather here can be severe, with high winds and fog. However, its isolation means there are large numbers of seals and sea lions. Santa Barbara, the smallest of the islands, lies between the main group and

Catalina. It was formed by volcanic activity and is the site of many seabird colonies.

For an alternative view of the Santa Barbara Islands, don't miss Catalina and its eclectic town, Avalon. Founded in 1887, Avalon is dominated by a large circular casino built on the waterfront by William Wrigley Jr, the chewing-gum magnate. Like some giant bastion it echoes the individual and laid-back atmosphere of this community of white wooden houses and turreted mansions.

Santa Cruz Island

Catalina is about more than just Avalon. There are a number of hiking trails with spectacular views from the ridges above the town. And a herd of bison roams the island's rugged hills – descendants of beasts brought in as extras for films shot here. It currently numbers some 250 animals and has been on Catalina since 1924.

Getting to the islands of the Channel Islands National Park and spending time on them is an adventure. They are accessible only by boat, and bad weather often results in cancellations. The outer islands – Santa Barbara, Santa Rosa and San Miguel – are reachable

Wild flowers clinging to the cliffs, Santa Cruz Island

Exploring the coastline, Santa Cruz Island

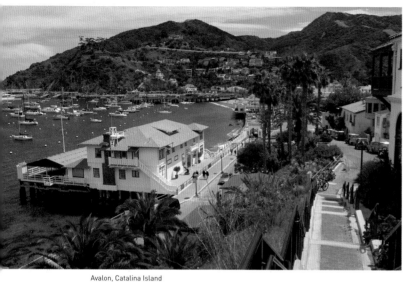

Avalon, Catalina Island

Green Pleasure Pier, Avalon, Catalina Island

Descanso Avenue, Avalon, Catalina Island

only in the best summer weather (with fewer than two hundred visitors a year making it to San Miguel). And overnight stays are possible only if you camp, which requires a permit from the National Parks Service. There are no facilities at the campsites, so you will have to take everything with you.

The islands' preservation is a relatively new development: they were incorporated as a national park in the 1980s. Previous exploitation has affected them but their ecosystem, if not pristine, is certainly unique and is recovering under the protection of the parks department.

Standing on Santa Cruz amid low meadows speckled with flowers, or on a high ridge with constant sea views, you could be lucky and spot migrating orca in the Santa Barbara Channel. What you will find most striking, however, is the contrast between the islands and the mainland of California, visible just 40 km – but a whole world – away.

ⓘ ··

There are daily boat services to Catalina from a number of ports, including Los Angeles. Island Packers has the sole concession for boat travel to the Channel Islands National Park from Oxnard and Ventura. Check the weather and schedules before departure as sailings are often cancelled or changed.

Green Pleasure Pier, Avalon, Catalina Island

Hong Kong Island
China

Looking down on the Victoria Harbour from the Peak

Hong Kong Island, dominated by the severe bulk of the Peak and with its craggy and thickly forested slopes littered with skyscrapers, is testament to man conquering nature. Developers here are masters at – and challenged by – taming the environment. The old colonial waterfront is now well back from the murky waters of Victoria Harbour, and land is constantly being reclaimed from the sea with new signature buildings rising to give Hong Kong a dynamic and fluid face.

The busy streets of Hong Kong Island

Britain forced China to cede the then undeveloped island in 1841, at the height of the first Opium War. In 1898 they leased it from China for 99 years, together with the Kowloon Peninsula and the New Territories on the mainland, and developed the colony of Hong Kong into one of the world's great financial centres. When the lease expired in 1997 it was handed back to the Chinese government. It was formed into a Special Administrative Region and its trade, growth and prosperity have continued unchanged.

Today Hong Kong Island is an alluring mixture of the new and the old. There seems to be no end to its architectural marvels – including the angular Bank of China Tower, which so offended adherents of feng shui. But enough of the old city has been preserved for this part to retain its original atmosphere. Stores in Wing Lok Street still sell

Skyscrapers on Hong Kong Island

Detail of the International Finance Centre on Hong Kong Island

Reflection of Jardine House

Star Ferry on Victoria Harbour

traditional remedies and birds' nests for soup, and in the Man Mo Temple massive incense coils hang from the ceiling, dispensing a fragrance that will live in your clothes for days. On the streets leading down from the Peak, small street markets sell everything from vegetables to fish, padlocks to batteries – and, of course, traditional fast food to workers hurrying to the banks and offices on the waterfront.

Asia Pacific Tower with the Bank of China building behind, seen from Hong Kong Park

Traditional Chinese medicine pharmacy

Signboards on Hong Kong Island

Star Ferry passing Hong Kong Island

The centrepiece of the old city is Statue Square, which dates back to the colonial era. The Legislative Council Building is the only one that survives from that period, but its granite bulk gives an idea of what the square – now dominated by the extravagant design of Norman Foster's HSBC Building – would have looked like.

One of the most unmissable sights in Hong Kong is the view from the Peak as the sun sets and the lights of the city come on. And at any time the scenery is breathtaking, although it can be obscured by a foggy haze created by pollution from the mainland. From here, the New Territories and Kowloon look closer than ever.

You can take the Peak tram to the summit or, to experience a world-class commute, go on the Mid-Levels Escalator – the longest moving staircase in the world. In the morning it runs down from the Peak, and in the afternoon it reverses. In a place where temperatures and humidity soar, the people of Hong Kong find it a useful way to get around. For another of the island's commutes – one that has to be the best in the world – board the double-decked Star Ferry, which will take even less time, and make the journey to the New Territories. The views are spectacular and the trip costs a couple of Hong Kong dollars.

Horse racing is a great passion in Hong Kong, and if you are in town on a Wednesday night there is a good chance that there will be a race meeting at the Happy Valley Racecourse. An institution in the city, it is always packed. You can buy a tourist ticket, which includes a meal in one of the private boxes on the top of the stands, or stand at the rail with the serious race-goers and see the action up close.

(i) ....

There are many international flights to Hong Kong. Visa rules are different from those for the rest of China, so check whether you need to get a visa before you leave. The classic place to stay on Hong Kong Island is the newly renovated Mandarin Oriental on Statue Square. Just a short walk from the Star Ferry, it has tremendous views across Victoria Harbour.

Happy Valley Racecourse

Giant incense coils at the Man Mo Temple

Amorgos
Greece

One of the Cycladic group of islands that rise out of the blue waters of the southern Aegean Sea, Amorgos is the Greek island you always dreamt of finding but never dared hope existed – or thought that, if it did, it would be long spoilt by the onslaught of package tourism.

Only 33 km long and no more than 6 km at its widest point, it is mountainous and hilly. Its peaks – the highest of which, Krikellos, reaches 823 metres – are often wreathed in fog or obscured by clouds that clear out to sea often as fast as they form. The island's three

Church of St Sophia, Langada

Nikouria Island, just off Amorgos

Fishing boat, Moúrou Bay

main settlements give you a taste of everything it has to offer: Katapola is the main port, Chora nestles atmospherically in the hills and Aegialis is a pretty seaside village.

Amorgos' most classically Greek feature has to be the monastery of Panagia Chozoviotissa. Situated 300 metres above the sea, it is halfway up a sheer cliff, under an overhang. The soaring, stark white edifice, reinforced with giant buttresses, is seemingly gouged out of the rock and is punctuated by tiny windows and topped by a triple bell tower.

The monastery of Panagia Chozoviotissa sits 300 metres up a cliff

The monastery of Panagia Chozoviotissa

Come here in the early morning, as the sun rises out of the Aegean and before the dozing cats in front of the small wooden main entrance have roused themselves. Knock loudly on the door and one of the three resident monks will invite you up, and, if you are lucky, share a piece of Turkish delight-style *luokoumi*, or even a shot of sweet *kitron* liqueur, with you.

The monastery was founded in 1088 after an icon of the Virgin Mary was miraculously brought to Amorgos from Palestine. It has changed over the years, influenced by the island's various rulers, including Venetians and Turks, but the icon is still presented to the faithful during a festival held on 21 November every year. Although the building is 40 metres long it is no more than 5 metres wide, with its eight storeys linked by a series of narrow rocky steps. There are a number of monks' cells as well as a treasury, a kitchen and storage rooms. The views from the monastery are breathtaking.

Another classically Greek attraction is Chora. This town consists almost entirely of white-walled houses – a feature of the Cyclades –

clinging to the hillside, and it has over 40 churches, including the smallest in the country with space for only three people.

Like most of Greece, Amorgos has a history that stretches back millennia. Above the port of Katapola lies the ruined city of Minoa, where you can discern a defensive wall and the outline of a gymnasium. There are also ruins of a Hellenistic temple from a later period.

There are a number of hiking trails around the island. One of the most popular leads from the monastery across the rugged interior to the village of Aegialis. From here, other trails take you to the traditional hill villages of Tholaria and Langada, with commanding views over Aegiali Bay.

The sea around Amorgos is the deep blue that is so associated with Greece, and Luc Besson filmed *The Big Blue*, his movie about

Shoreline by Panagia Chozoviotissa monastery

Entrance to Panagia Chozoviotissa monastery

AMORGOS

Aghios Ioannis Theologos (St John the Theologos)

View of Nikouria Island from Aghios Ioannis

Herding goats

free divers, here in 1988. Although it put the island on the international map, it luckily didn't lead to the influx of tourists that has blighted many other film locations around the world.

ⓘ ....

Amorgos is on the south-eastern edge of the Cyclades. There are regular ferries to the island from Piraeus and Naxos, both of which can be reached by air from Athens and possibly by charter flights from other European cities in the summer season. Although Amorgos is small, a rental car is useful to help you get around. The Aegialis Hotel in the town of Aegialis is the best on the island and has wonderful views over the bay.

Tholaria, Amorgos

The Golden Temple
Amritsar, India

Detail on the roof of the Sri Harimandir Sahib

Golden temple reflected in the Amrit Sarovar

The Sri Harimandir Sahib, or Golden Temple, arguably the most beautiful building in India, is set on an island in the middle of the Amrit Sarovar (Pool of Nectar), the lake that gives the town of Amritsar its name.

Construction of the temple, the holiest building of the Sikh religion, started in 1589, after someone who was disabled had reputedly been cured by bathing in what was then a small lake. The lake was expanded to form the Pool of Nectar. The land for this enlargement was granted by the Mogul emperor Akbar. The original temple was destroyed by invading Afgans in 1757 and the site had to be retaken by the Sikhs. It

was rebuilt in 1764, and in 1830 its roof was covered with 100 kilograms of gold – the reason why the building is known as the Golden Temple. Up close, the detail of the gilding is exquisite.

As the spiritual centre of the Sikh faith the temple is about far more than beauty. Sikhism is noted for its egalitarianism and anyone is welcome here, irrespective of religion, race or caste. The overriding impression is one of peace and spirituality, despite the crowds of visitors. Every Sikh tries to make the pilgrimage to the Golden Temple at least once, and the atmosphere is surprisingly cosmopolitan. Doctors from the US mingle with fierce old men from rural villages.

As at other Sikh temples (*gurudwara*) there is no organized worship. Some pilgrims stroll around the Pool of Nectar. Others sit and stare at the Golden Temple in quiet contemplation. Many bathe, lowering themselves into the water with the help of great chains anchored to the sides of the lake.

Looking down the causeway

The inside of the temple is even more beautiful than its exterior, with breathtaking detail and a pietra dura inlay of precious stones in marble that is even finer than that in the Taj Mahal. The place is imbued with an air of deep devotion. In the main sanctum a priest surrounded by hundreds of quietly sitting pilgrims reads from the *Granth Sahib*, the holy book of the Sikh religion. A complete reading takes 48 hours. The atmosphere is incredibly moving, yet humblingly simple. The feelings of reverence and awe are palpable. Another reading of the *Granth Sahib* takes place upstairs, and yet another in a chapel on the roof. Early in the morning the book is ceremonially processed from the Akal Takht, where it is kept, and returned late at night.

The Golden Temple complex

Temple guard contemplates the Sri Harimandir Sahib

Inside the Sri Harimandir Sahib

Handing out plates at the Guru Ram Das Langar

The Pool of Nectar is a special place for sitting and contemplating, but make sure you don't point your feet at the temple or dip them in the water. Doing either is considered to be an insult, and you certainly don't want to insult Sikhs. They are proud people with warrior traditions – as is shown by the wall plaques that commemorate army detachments. A priest or guard may fix you with a fierce gaze, looking like thunder – then smile, like a rainbow breaking across clouds.

No one could be more welcoming than a Sikh. Hospitality is central to Sikhism and you may well be offered free cups of tea. You will certainly be offered free meals. The Guru Ram Das Langar is an enormous kitchen and dining hall where vegetarian food is provided for pilgrims. Everyone sits in long lines on the floor of the hall holding metal trays on to which their portions are put. Extra chapattis are offered – just hold your hands together and one will be dropped into them. At the back of the dining hall is the kitchen, where chapattis are made in a remarkable production line and volunteers wash up the trays.

The peace and spotlessness of the temple complex contrast markedly with the bustling, dusty and run-down old town that surrounds it. Narrow bumpy streets are thronged with garrulous cycle-rickshaw drivers and crowds of pilgrims. It would be a shame

to rush through this area, though, as there are a number of fascinating shops with everything a devout Sikh could ever need: religious
pennants, turbans, bracelets and even martial swords.

ⓘ ···

Jet Airways fly directly to Amritsar from London. Most of the better hotels are
some way out of town. Local arrangements can be made through Atithi Travels
in Delhi. Although visitors are welcome inside this holiest place in the Sikh faith,
all photography is banned. We obtained special permission from the temple
authorities to take the pictures shown here.

Man reading at an upper window of Sri Harimandir Sahib

Golden Temple floodlit at night

Fishing marina, Key West

The Florida Keys are a unique chain of 1700 islands that stretch 177 km into the Atlantic Ocean from mainland Florida and are linked by the Overseas Highway. One of the world's great drives, this takes you over a series of 40 bridges that skim the blues of the Atlantic on one side and the aquamarines of the Gulf of Mexico and Florida Bay on the other.

In its strict sense, a key is an ancient coral reef raised above sea level. The Florida Keys are split into three groups – Upper, Middle and Lower – whose characters change the further you get from the mainland.

In the Upper Keys the first main island is Key Largo, made famous by Humphrey Bogart and Lauren Bacall in the gangster film of the same name. Continuing the Bogart theme, you can visit the boat that was used in the 1951 movie *The African Queen*. The next notable

The Florida Keys

USA

Sunset, Key Largo

stopping point in this part of the Florida Keys is a group of islands called Islamorada (Purple Isle). It is thought that Spanish explorers named them after floating purple sea snails. Islamorada bills itself as the Sportfishing Capital of the World.

Long Key is one of the Middle Keys, and in its state park you can walk through the sort of untouched tropical forest that covered much of the keys as little as a hundred years ago.

About halfway down the Florida Keys chain, and spanning a number of separate islands, is the small town of Marathon. Its name is said to come from the seemingly 'marathon' workload of the men

Long Key State Park, Long Key

Islamorada

Papa Joe's Restaurant, Islamorada

A myopic local, Key West

who built the Florida East Coast Railway that once ran the entire length of the keys. Here you will find the Crane Point Hammock Reserve, which has more Pre-Columbian archaeological remains than anywhere else in the keys.

Like Long Key, Bahia Honda in the Lower Keys is the site of a state park. Its beach is a 4-km sweep of white sand and is reputed to be one of the most beautiful in the United States.

The Seven Mile Bridge spans the Middle and Lower Keys, and is among the longest bridges in the world. It has featured in countless films and crossing it is probably the highlight of the whole drive. In the middle it rises 20 metres above the water to allow boats to pass under it.

The last island in the chain, Key West, is closer to Cuba than it is to the United States mainland. This gives it a unique atmosphere, and makes it a haven for anyone seeking an alternative way of life. Its reputation for 'living for today' has attracted writers and artists for generations. Of these, the one who is most synonymous with Key West is Ernest Hemingway, who lived and worked here for some of his most prolific years. His house on Whitehead Street is now a museum, and the descendants of his six-toed cats still wander the grounds. When he wasn't writing (or drinking and fighting) Hemingway used to fish for blue marlin and sailfish in the Florida Straits between the island and Cuba.

A typical Key West house

Sloppy Joe's Bar, Duval Street, Key West

Key West is very much at the edge in the meteorological as well as the geographical sense. With nowhere for people to go but 200 km back up the Overseas Highway, each hurricane season brings with it the potential for disaster, which is dismissed with studied nonchalance by the local population, who call themselves 'Conchs'.

The Everglades, a vast wetland of subtropical marshes on the coast of the mainland, provides an interesting contrast to the Florida Keys. Reachable from the top of the Overseas Highway, its southern part is incorporated as the Everglades National Park. It is most famous for its alligator population, but the area is also home to a vast array of birdlife.

ⓘ ···

The Florida Keys are easily reached by driving from Miami, which has some of the best international flight connections of any city in the United States. The hurricane season is from June to early November, and is a good time to avoid the keys, even though they won't necessarily be hit by hurricanes!

Bali

Indonesia

Sunrise at Gunung Batur

The enduring image of Bali is one of luminous-green rice terraces, topped by often cloud-wreathed volcanoes. Lushly extravagant, the terraces appear to score every hill and seemingly no space is wasted; they hug every contour and some of the cultivated areas are only a few feet wide.

The dominant culture of Bali is a unique blend of Hinduism and animism. The worship of Hindu gods and adoration of the epic heroes of the *Ramayana* are combined with a belief in spirits that leads each

village to have three temples: the *pura puseh* for the community's founders; the *pura desa* or village temple; and the *pura dalem* or temple of the dead. Most houses also have a spirit house where daily offerings are made to placate the occupants' ancestors.

Many of the temples seem to be permanently deserted, but each hosts an annual festival when local people wearing traditional dress bring ornate offerings of fruit and other goods. There is a bewildering calendar of festivals in Bali, including an extended New Year period every 210 days.

Mythical creature, Ubud

Paddy fields near Ubud

Monkey Forest, Ubud

There are obligations on a personal level as well. Births and their anniversaries, weddings and cremations, all demand involved and often costly rituals and offerings. Much of the income of the average Balinese is taken up with observing such traditions.

Bali has been a popular tourist destination for generations, and although it has changed in many ways its strong culture, which permeates all levels of society, has helped it to retain its essential nature. Balinese dance is central to this and can often be seen in processions and at temple festivals, while professional dancers give regular and spirited displays, especially in and around Ubud, the undisputed cultural capital of the island.

Many of the dances are based on Balinese interpretations of the great Hindu epics. The *Ramayana* ballet, in the atmospheric City Palace in central Ubud, is the most traditional, but the Barong features a mythical lion – you can often see these creatures in temple processions. One of the most enigmatic dances, the Kechak, depicts an episode in the *Ramayana* featuring the monkey god Hanuman. Dozens of dancers act as his monkey warriors, synchronizing rhythmic movements and chants.

Monkeys are regarded as holy in Hinduism, and you should make a visit to the Monkey Forest and atmospheric temple of the dead on the outskirts of Ubud. The ancient temple is surrounded by an ancient forest, which is home to troupes of rather assertive monkeys.

Over the years Ubud has grown from a small village into a town with many guest houses and boutiques, but its essential spirit is the same as it was some 20 years ago. Despite all the commercialism, the culture seems to fill its streets. There are fewer paddy fields ringing it than there used to be, but you can still stroll down a small alley in the middle of town and come across a hidden valley bedecked with rice terraces.

One of the most beautiful sights on Bali is the sun rising over the volcano of Gunung Batur and the crater lake beside it. If you make a

Temple festival for Balinese New Year

very early start it is possible to climb the volcano, but the views are equally good from the lip of the crater, and the volcano and lake will be in the same panorama. The sight is breathtaking – especially when morning mist from the lake fills the crater and glows, backlit under a vivid scarlet sky.

Flooded rice paddies near Ubud

It would be impossible to write about Bali and ignore the two terrible terrorist attacks. Certainly there is a risk, as there is in so many destinations around the world, but few places have managed to retain such cultural integrity in the face of so many challenges.

ⓘ ··

It is surprisingly difficult to get to Bali from many places in the world. There are direct flights from Australia and from the Indonesian capital, Jakarta. Many budget airlines fly to Jakarta from other South-east Asian cities. The Aman chain has stylishly exclusive properties in Ubud, Candi Dasa and Nusa Dua. Although it offers a package that includes transfers between resorts, hiring a car will enable you to explore by yourself. Bali's main temple complex is at Besakih, but the buildings are closed to tourists and the hassle level there is so high that it is best avoided.

Lava flow in front of Gunung Batur volcano

Franciscan Monastery, Vis Town, Vis

The Croatian coast is littered with beautiful and historic islands, and choosing which ones to visit can be a Herculean task. Some of the most fascinating – and least known to tourists – are those of southern Dalmatia, including Vis, Lastovo and Korčula.

Looking down to the harbour of Komiža, Vis

The Southern Dalmatian Islands can be visited from the coastal town of Split; and, although there are regular ferry schedules, the best way to explore what are some of the most untouched corners in the Mediterranean is to charter your own sailing yacht for a week. This will give you access to hidden coves with stunning azure waters, tiny marinas virtually unreachable from the land and ancient harbours steeped in history.

The island of Vis has two main harbours with wildly different characters. The waterfront at Komiža has 16th-century Venetian-style houses in the middle and a severe-looking fortress guarding the entrance. Craggy hillsides dominate the town and provide a rugged backdrop as you sail in. A fortified 17th-century Benedictine monastery

Benedictine monastery overlooking Komiža Harbour

on a small hill nearby gives commanding views at sunrise. Komiža feels like a forgotten and ancient shelter from the sea. The town of Vis, on the other side of the island, is far more cosmopolitan. A long waterfront with a distinctly French atmosphere looks out on to a narrow peninsula, the site of a Franciscan monastery.

The island is well worth exploring as there are a number of tiny coves and simple farms, as well as amazing views out to sea. Vis was briefly the headquarters for Tito's partisans in 1944, and was a military zone closed to foreigners as recently as 1989.

Having your own boat means you will be able to sail to the neighbouring island of Biševo, which is famous for its Blue Cave – so called because the sun shining through the first, submarine, entrance to the enclosed cave creates an aquamarine light. Small boats ferry you through a seemingly impossibly small second entrance and through the gloom until you are bathed in iridescent blue.

Lastovo is one of the most untouched of the outer Southern Dalmatian Islands. It has few tourist facilities, but in the tiny natural harbour of Zaklopatica an enterprising restaurateur will allow you to

Korčula Cathedral

Milna, Brač Island

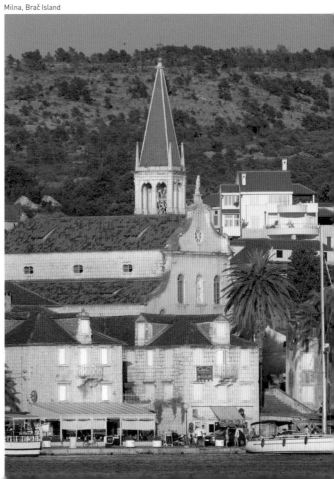

moor your yacht outside his restaurant provided you eat inside. The order of the day is seafood – some of the best I've ever tasted, and certainly the freshest. The menu depends on what has been caught by the fishermen of the hamlet. When I arrived they had just landed a monster tuna, which formed the basis for many dishes.

Lastovo town lies on a steep hill 3 km from Zaklopatica, in a natural amphitheatre that faces away from the sea. A network of 15th- and 16th-century houses with characteristic towered chimneys jumbles around a maze of narrow alleys that scatter down the hillside. The town is topped by a church and a fort.

Steep coastline of Brač Island

The ancient walled town of Korčula on the island of the same name is firmly on the tourist trail as the birthplace of Marco Polo; one of the world's great explorers he will have seen many of the islands in this book. It is humbling to walk its fortifications and imagine how he might have looked out to sea when he was young, dreaming of adventures to come. The well-worn flagstones glint like molten gold as the setting sun sends fingers of light through the narrow alleyways that lead down to the sea.

There are many historical sights to see in the islands, but the beauty of having your own yacht is that it gives you the freedom to explore places that are well away from tourist destinations. You can drop anchor in a hidden cove of turquoise water and swim before preparing lunch on board. In the evening you can simply moor alongside the waterfront of an ancient harbour and wander to a local restaurant, before returning to a room that must have the best view in town.

ⓘ ··

Chartering a yacht is more affordable than you might have thought. Sail Croatia has a number of boats that sleep up to eight people. They will also supply a skipper who will help you with the route and, of course, sail the boat! They have offices in London and around Croatia, including Split. Croatia Airways has indirect scheduled flights to Split, but you will probably find a direct charter quicker and cheaper during the summer months. Boat charters run from Saturday to Saturday.

Boat moored in Zaklopatica harbour, Lastovo

Yasawa Islands
Fiji

Overlooking the Yasawa chain from a seaplane above Nanuyalailai Island

The Yasawa Islands have a brooding malevolence that is a complete contrast to the relaxed and easy-going nature of their inhabitants. Black rocks pepper soft golden beaches, and craggy green hillsides covered in ominously dark grassland are rent with valleys and ridges that show the violence of their volcanic creation.

The people of Fiji have not always been friendly. Historically, they have fought with each other and any outsiders, and built up a reputation for warlike behaviour. Although the islands were briefly

Rugged interior of Nacula Island

visited by the explorers Abel Tasman, in the 17th century, and Captain Cook in the 18th, they were largely avoided by Europeans until the discovery of valuable sandalwood in 1804.

Trade led to settlers and missionaries coming to the islands from Europe, but this exacerbated the conflict between the Fijian chiefdoms, and the country was eventually ceded to Britain in 1874. The British introduced Indian labourers, whose descendants now make up almost half the population, and ruled Fiji until it gained its independence in 1970.

The Yasawa are a string of some 20 volcanic islands that stretch in an 80-km line in the west of the country. The water between Nanuyalailai, Nanuya, Matacawalevu and Tavewa is particularly clear and in the sunlight it glows an iridescent blue – so blue that the 1980 Brooke Shields film, *The Blue Lagoon*, was shot here. The lagoon is now universally known by the name of the movie, and is fantastic for diving and snorkelling – the water so clear that visibility is near perfect.

Beach on the island of Tavewa

Nacula Island

What I particularly like about the Yasawa Islands is that the accommodation includes a number of stylish mid-range and even backpacker places, as well as a couple of super-expensive ones. This means you don't have to be rich to swim in the fabled Blue Lagoon, or even stay on the beach.

The looming height of many of the islands means there are some perfect vantage points if you are prepared to climb. Getting to the top of Tavewa is a bit of a slog, but the views over the Blue Lagoon are amazing. Nacula provides even more spectacular scenery. Stand on a high ridge and you will be rewarded by a 360-degree panorama, with the lagoon and the islands stretching away in the distance on one side, and the harshly weathered valleys and inlets of Nacula on the other. The combination of colours – dark green hills running down into dark blue waters fringed by golden sand – is outstanding.

Blue lagoon off Tavewa Island

There are villages on most of the islands, and many of them are linked to guest houses and resorts. This gives the Yasawas a strong feeling of culture and means the local people benefit from tourism. Almost all the houses are made from modern materials, but you will still see some traditional thatched huts. Instead of being centred around their chief's house, most villages are now focused on churches, but a chief is still treated with great respect and runs the

Boat passing Matacawalevu Island at sunset from Nanuyalailai Island community. It is possible to visit the villages, but you should take a guide – and kava for the chief.

Kava (*yaqona* in Fijian) is a traditional and bitter-tasting drink made from a root. It is mildly intoxicating and used in a number of ceremonies. Village elders drink it before making decisions and it is

Volcanic rock of Nanuyalailai Island with Nacula behind

Palm fronds outside a traditional hut, Nanuyalailai Island

often offered to guests. Sampling *yaqona* is as much a part of a visit to the Yasawa Islands as lying on the beach or swimming in the Blue Lagoon – although it is drunk more for its effect, and for cultural reasons, than for its taste!

ⓘ ···

Fiji's main island, Suva, is easily reached from Australia or New Zealand. Air Vanuatu also has regular flights from Port Vila, allowing you to combine a visit to Fiji with one to Vanuatu. From Suva a fast ferry, the *Yasawa Flyer*, stops off daily at all the main Yasawa island groups. There is also a small four-seater seaplane, run by Turtle Airways. The family-owned and run Nanuya Island Resort is right on the Blue Lagoon and is a perfect blend of style and affordability. There is a PADI dive centre and the food in the open-sided restaurant is excellent.

Madagascar is known the world over for its unique wildlife. It split from the east coast of Africa millions of years ago and its flora and fauna developed in isolation, producing many endemic species – 80 per cent of which are found only on the island.

'Red Tsingy' near Diego Suarez

The signature animal of Madagascar has to be the lemur, a member of the primate family. It comes in all shapes and sizes, from the large sifaka, which leaps along the ground between trees on its hind legs, to the tiny aye-aye, which is nocturnal and lives in the hollows of trees. Because it uses its abnormally long middle finger to kill its prey, local people believe it is a bringer of death, and many are killed on sight.

The capital of Madagascar, Antananarivo is in many ways more like a provincial French town than an African city. Centred on a long rocky ridge, it is topped by two cathedrals, government buildings and even the remains of a royal palace that was set on fire in 1995. Spreading down from the ridge is a jumble of ancient streets and houses.

Baobab at sunset, near Morondova

Baobab reflected in a paddy field near Morondova

Zebu cart passing the Avenue of Baobabs, Morondova

Deeply rooted beliefs are part of the culture of the Malagasy of Madagascar, who are predominantly a mix of Asian and African peoples. For instance, chameleons are said to be the spirits of dead ancestors, and many local people go out of their way to avoid hurting one – even when they are driving. There is a wide variety of species on the island including a minuscule chameleon less than a centimetre long. Because their eyes rotate, allowing them to see in all directions, it is believed that chameleons are able to see into the future and the past.

The landscape of Madagascar is striking. The island is noted for its *tsingy*, the Malagasy name for a bizarre, spiky karst limestone rock formation caused by erosion. Many of the spikes are razor sharp and there are great crevices, some of which allow trees and bushes to grow.

The *tsingy* include one in the Ankarana Reserve in the north, near the Diego Suarez, and the Tsingy de Bemaraha in the south. They often form vast fields that stretch as far as the horizon and create impenetrable barriers. Any vegetation that manages to find a foothold struggles to thrive in this hostile environment.

There are six endemic species of baobab in Madagascar, and the Morondova region is where to find them – as your plane comes in to land you will see what seem to be thousands of the trees littering the countryside. A visit to the Kirindy Reserve in search of lemurs will take you through the Avenue of Baobabs – a dirt road that leads through a small village and is lined with these great trees. The ones here are particularly tall and thin, and tower over the local people.

Tsingy at Ankarana Reserve

Antananarivo

Verreaux's sifaka lemur (*Propithecus verreauxi*)

Male common chameleon (*Chamaeleo oustaleti*)

Crowned lemur (*Eulemur coronatus*)

Large day gecko (*Phelsuma madagascariensis*)

If the lemur is the signature wild animal of Madagascar, the signature domesticated one must be the humpbacked zebu, an animal that is much prized for its milk and meat, and even for transport. As the baobab avenue glows a golden red at sunset, carts drawn by zebu clatter slowly through the giant trees, which cast their long shadows over villagers returning home at the end of a day spent working in the fields.

ⓘ ....

Air Madagascar flies to the capital, Antananarivo, from Paris, Milan and Johannesburg among other places. It also has an extensive network of internal flights to Madagascar's main cities. Le Voyageur, a Swiss-run, Madagascan-based company, can make local arrangements and organize tours around the island.

Avenue of Baobabs, Morondova

Countryside near Diego Suarez

Stockholm
Sweden

Stockholm is built on 14 islands that lie between Lake Mälaren and the Baltic Sea and are part of Sweden's most extensive archipelago. And the heart of the city, the 13th-century Gamla Stan or medieval old town, spreads over three of them.

This part of Stockholm is a meandering warren of picturesque streets and alleys lined with towering red, orange and yellow façades. If you wander around you are virtually guaranteed to get pleasantly lost. Its biggest attraction is the huge Royal Palace (Kungliga Slottet), the official residence of the Swedish royal family. It has been rebuilt a

Riddarholmen (left) and Gamla Stan seen from Södermalm

Royal Palace

number of times, most recently following a fire in the 17th century. The current building dates from 1754, although work on it continued for another 80 years. The Swedes have learnt to make the most of their short summer, before the darkness of the northern winter closes in, and this is the perfect time to visit Stockholm. People take advantage of the long hours of daylight to enjoy late-night visits to the city's parks and outdoor spaces. Work hours are relaxed, there are free festivals and concerts, and alfresco bars and cafés are packed.

June, July and August are also the best months for exploring the archipelago that sprawls to the east of the city. The islands – more

than 24,000 cover an area of 150 km by 80 km – are reckoned to be 2 billion years old and are low and rounded, shaped by the smoothing effect of glaciers during the last ice age. The ice affected them in another way too: its massive weight pushed them down into the sea, and now that they have been released they are rising by about 4 mm a year.

Sailing is immensely popular in Sweden and the waters around the archipelago are treated like a lake – even though they are actually part of the Baltic Sea. Because it is fed by many rivers, and only

Östermalm seen from Skeppsholmen

Riddarholmskyrkan

meets other oceans at two points, the Baltic has a relatively low salt content and often freezes in the winter.

The network of channels in the archipelago forms an intriguing maze that is perfect for island hopping by sail, motor boat or ferry. The shorelines of the islands, with their bays and inlets, are typically covered with coniferous and deciduous trees, peppered with worn rocks and boulders and studded with stocky wood cabins and boat sheds.

There are some 50,000 holiday homes in the archipelago and the islands near Stockholm can be crowded at weekends. To get a taste

Ferry from Nämdö Böte at sunset

of the true isolated beauty of the more outlying ones it is worth heading for Nämdö – 40 km and a whole world away from the city centre. It has 35 permanent residents, a single dirt road, a restaurant, a grocery store and a school with seven pupils. Amid its picturesque, moss-covered forests and wildflower-speckled meadows there are a number of hiking trails where you will be sure to see roe and fallow deer – even elk, which often swim here from other islands. At the northern end of Nämdö there is a high vantage point, complete with a

Östanvik, Nämdö

Sunset seen from Nämdö Böte

tower that affords a 360-degree view over the surrounding archipelago – perfect as the setting sun turns both sky and water a vivid gold.

ⓘ ··

Scandinavian Airlines (SAS) flies to Stockholm from many countries. The world's largest fleet of steamships operates between the city and the islands of the archipelago. Ferries are run by Waxholmsbolaget. Schedules vary with the season. For Nämdö, take a bus from Stockholm to Stavsnäs, then the one-hour ferry ride.

Lord Howe Island
Australia

Lord Howe Island is 550 km from the coast of New South Wales and its unique ecology owes much to this isolation. Thrown up by a massive volcanic eruption on the seabed millions of years ago, it has developed many new species and is fringed by the world's southernmost coral reef. Half of the island's plant species are endemic, including the Kentia palms, which are grown here and exported as houseplants. There is one endemic bird (the Lord Howe woodhen) and even endemic glowing mushrooms that only appear after heavy rain.

The island was discovered in 1778 and named after Admiral Lord Howe, the First Lord of the Admiralty. It was 55 years before the first

Looking from the summit of Mount Gower to Mount Lidgbird

The lagoon and Mounts Gower and Lidgbird

Rainbow sweeping across the lagoon

settlers arrived. A second group followed a few years later, and many of the 340 people currently living here trace their ancestry back to them.

Lord Howe has a uniquely laid-back feel. The main settlement is tiny with a small selection of shops, a couple of restaurants, a bank, post office and a community centre that doubles as a cinema. Everyone knows each other and the pace of life is languid, exemplified by the 25 kph speed limit for the island's few vehicles – although most people make do with bicycles.

Only 11 km long and no more than 3 km wide at any one point, Lord Howe curls round the shallow lagoon that is home to its coral

The approach to Mount Gower

Scaling Mount Gower

reef, and is dominated by the twin mountains of Gower and Lidgbird. The climb up the former is one of the great challenges of the island: at 875 metres above sea level it is nearly 100 metres higher than Lidgbird. You will be escorted by local guide Jack Shick, and although there is no technical climbing the expedition is not for the unfit or anyone who suffers from vertigo.

After a steep climb up a thickly forested hillside the first challenge is to traverse the slopes of Mount Lidgbird, on a narrow path across vertical cliffs that tower above you and drop to the sea. Thankfully, there is a rope! From here it is basically uphill all the way to a sharp ridge between the two mountains, then a vertical scramble up a rock face to the summit of Mount Gower. In many places you will have to pull yourself up on ropes. The slopes are largely covered with stunted trees and bushes, so you won't be aware of the risks involved in this part of the climb.

At the top of the mountain you will find a flat plateau covered with a misty forest of giant ferns and stunted trees. Moss and lichen seem to cover every surface and it truly feels like a different world. The summit is often wreathed in cloud, and there are even vicious

Secluded beaches on the lagoon

Muttonbird on the summit of Mount Gower

rainstorms – the weather on Lord Howe is changeable to say the least – and the plateau has developed its own ecology. The pumpkin tree you will see here is found nowhere else on earth.

At a call from Jack, ungainly short-tailed shearwater (colloquially known as muttonbirds) will come crashing through the trees to land with a thump. They will investigate you with great interest before they become distracted and run around pecking each other chaotically.

The views from the top of Gower, across the island and to the high rock-stack of Balls Pyramid 23 km out to sea, are fantastic, but your enjoyment of them will be tinged with the realization that you still have to make your way back down.

There are equally stunning vistas from the Malabar Hill on the north end of Lord Howe. At only 209 metres, reaching its summit is a

Driftwood on the beach

mere stroll compared to climbing Gower, but you will enjoy commanding views of that mountain. Between September and May you can see countless red-tailed tropic birds nesting.

Lord Howe is home to a number of beautiful beaches, including Ned's Beach, a near-perfect cove. Giant yellowtail kingfish that have got used to being fed by hand linger here, amid the crashing surf, and will swim expectantly up to you.

ⓘ ··

Qantaslink fly to Lord Howe Island from Sydney daily and from Brisbane at weekends. Accommodation must be organized before you fly. The Capella Lodge is a boutique hotel with great food, and overlooks the lagoon, as well as the twin mountains of Lidgbird and Gower. The hotel has a spa, which you will definitely need if you attempt to climb Mount Gower!

Plantlife on the summit of Mount Gower

Giant Buddha, Dambulla

Sri Lanka has to be one of the world's most varied islands. Individually, its history, culture, wildlife or stunning beaches would justify a visit in their own right; together they make travelling through the island a unique experience.

Temple of the Tooth, Kandy

The very best of Sri Lanka can be seen in a simple loop from the international airport at Colombo. Within just a few hours you reach Sigiriya and its famous palace on top of an enormous rock – a hardened plug of magma from an ancient eroded volcano – that towers some 200 metres above the surrounding plain.

The Singhalese king Kasyapa built the palace at the end of the 5th century AD on what was formerly the site of a Buddhist monastery. The climb to see its ruins is steep and not for the faint-hearted, especially if you take a detour about halfway up the cast-iron spiral staircase to see the exquisite frescoes of scantily clad women under an overhang. Once you arrive at the top the views – as far as the hills around Kandy – are spectacular.

Sigiriya is a good base from which to explore Anuradapura and Polonnaruwa, the island's deserted former capitals. Both have a number of Buddhist relics, including stupas and giant images of the

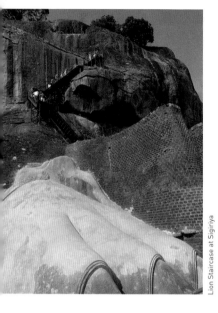

Lion Staircase at Sigiriya

Buddha, and Polonnaruwa is home to the iconic reclining Buddha seen in so many Sri Lankan tourist publications. It is also home to a number of particularly cheeky langur monkeys.

Sri Lanka has a relatively large population of wild elephants, many of which seemingly wander around the countryside; migration corridors link a number of national parks. Minneriya, a short distance from Sigiriya, is set around an ancient man-made reservoir, a permanent source of water, and in the dry season it can attract hundreds of elephants, making it one of the best parks for seeing these animals.

Kandy is a few hours' drive south of Sigiriya, and on your way there it is well worth visiting the fascinating series of caves at Dambulla. These house ornate frescoes and ancient images of the Buddha, some of which date back as far as the 1st century BC.

Reclining Buddha, Dambulla

Drummer at the Temple of the Tooth, Kandy

Tea plantation and waterfall, Nuwara Eliya

The city of Kandy is noted for the the Temple of the Tooth (Sri Dalada Maligawa), which houses the most sacred relic in Sri Lanka: a tooth of the Lord Buddha. The golden casket that encases it is presented to the faithful three times a day, in a crowded and somewhat noisy ritual. To the accompaniment of frenetic drumming and horn playing, a long line of pilgrims and tourists files past the open chapel in which the tooth is kept. Kandy is also known for its sprawling botanical gardens, which date back to the era when Britain ruled the island – then known as Ceylon.

Elephants at Minneriya National Park

Sri Lanka is famous for its tea, still marketed as Ceylon Tea. The plants were brought here from India by the British when their attempts to grow coffee failed, and the centre of tea production is Nuwara Eliya, about three hours' drive south of Kandy and some 2000 metres above sea level. The slopes of all the hills around and leading up to the town are covered in tea plants, and are often worked into intricate patterns by the tea pickers as they move around. The new leaves of this remarkable plant are harvested by hand when they are a month old by legions of women who work six days a week, often on vertiginous hillsides.

Botanical Gardens, Kandy

Stormy coastline, Galle

If you have always hankered after seeing leopards, Yala National Park on Sri Lanka's south coast (about five hours' drive from Nuwara Eliya) must be one of the best places in the world to go. Not only is the population here high, but as the top predators these big cats aren't as elusive as they are in Africa. Yala is also home to a good many elephants and a number of bears.

As you drive back to Colombo you will pass the historic coastal town of Galle. The area was among those worst hit by the Boxing Day tsunami in 2004. Thousands were killed, although many people in the old town, protected by the walls of the 1663 Dutch fort, survived. The coastline around Galle boasts some of the most beautiful beaches on the island, many of which are deserted, with little or no development.

ⓘ ....

Emirates flies to Colombo from a number of UK airports, with a connecting flight in Dubai. Public transport is somewhat unpredictable in Sri Lanka, but hiring a car and driver is good value. Jetwing Holidays can arrange this for you, and you will stay at a number of their properties, including the stylish Vil Uyana in Sigiriya, the historic St Andrews in Nuwara Eliya and the luxurious lighthouse in Galle. Their Kandy property is at the stunning Hunas Falls, 32 km from the city. If you want to be closer Jetwing suggest the Citadel. Their property in Yala was destroyed by the 2004 tsunami so they recommend the Yala Village.

Painted stork and deer at Yala National Park

Sark

Channel Islands, United Kingdom

Sark is the smallest of the main group of Channel Islands between England and France, and arguably the most distinctive. Its age-old political system – it is governed by the *seigneur* and a parliament called the Chief Pleas – means it has the last feudal constitution in the western world.

Although Sark was granted in perpetuity to Helier de Carteret, the first *seigneur*, by Elizabeth I in 1565, it is part of the British Isles and owned by the British Crown. It is quaintly English but with a number

La Coupé leading to Little Sark

Lighthouse

of unique quirks, the first of which becomes apparent when you arrive on the ferry from Guernsey or Jersey. Save for tractors that transport supplies from the tiny harbour and take tourist bags to visitors' accommodation there is no motorized transport. You simply tell the baggage handler where you are staying and your luggage arrives in a couple of hours. There are a number of horse-drawn carriages that take tourists on sightseeing tours, but these are officially discouraged from working on Sundays. Alternatively, for more run-of-the mill transport you can hire a bicycle.

Cliff-top heather

Tractor parking by the pub

Although there is little crime on Sark, the island has a couple of part-time police officers who are elected from the community, and a tiny two-cell prison that dates back to 1856. Anyone who has been arrested for a minor crime such as vagrancy or drunkenness can be imprisoned for up to two days, simply at the behest of the officers. More serious cases are referred to Guernsey. The prison has a distinctly Victorian feel, and is still used occasionally.

Sark consists of two separate islands, Big Sark and Little Sark, linked by a vertiginous causeway known as La Coupé, built in 1945 by German prisoners of war. Although it has railings cyclists must dismount and wheel their bicycles over it, while passengers in carriages have to get out and walk. On one side a cliff drops sharply to Convanche Bay 300 metres below; on the other, the slope down to Grand Grêve, one of the most popular sandy beaches on the island, is less forbidding.

Much of Sark consists of rolling farmland that leads to steep cliffs; the edges of many of these are exposed and covered with bracken and heather. There are few roads, and those that do exist often run between high banks topped with hedgerows. In some places they are lined with trees whose branches meet to create shady canopies.

The small main settlement, known as the Village, is just up from the harbour. Its high street, the Avenue, has a few shops and cafés, and a post office with the only postbox on the island – painted blue and not red. Many of the houses in the neighbouring streets are at least a couple of centuries old and built from local stone.

The Seigneurie, the official residence of the *seigneur*, was built in 1675 on the site of a 6th-century monastery to the north of the Village, but there have been additions over the years – notably a large tower added in the Victorian era. Its beautiful formal gardens are open to the public and are one of the most popular attractions on Sark.

Hay bales

La Seigneurie and gardens

There are a number of walks around the island. Alternatively, you can take one of the boat trips around Sark run by George Guille. His ancestors were among the original Sarkese who came here in 1565, and after the trip he will happily regale you with tales about the island in the Bel Air Inn, one of its traditional English pubs.

La Sablonnerie on Little Sark

Stone house in the Village

ⓘ ....

Sark can be reached by regular ferries from Guernsey and Jersey. Aurigny Air Services links these islands with a number of airports in the United Kingdom, including London Gatwick. It also flies to Dinard in France, which allows you to combine a visit to Sark with one to Mont Saint-Michel (see page 174).

Carriage crossing La Coupé

Bacuit Archipelago
Palawan, Philippines

Beach at Seven Commandos Point on the Palawan mainland

The island province of Palawan is unique in the Philippines. It lies on the Eurasian tectonic plate, whereas the other islands in the country are on the volcanic Philippine plate. The result is a landscape that is unlike anywhere else in the vast Filipino archipelago.

The north-west tip of Palawan Island itself, and the 45 islands within Bacuit Bay, have been incorporated by the Philippine government as the El Nido Protected Area – a paradise of hidden bays, coves and lagoons.

The landscape in the reserve is formed from grey karst limestone, and over 250 million years' erosion from rainfall has sculpted this into fantastic shapes that resemble dripping wax more than rock. Jagged cliffs and ridges vie with fantastic vertical spires. Sharp, and difficult to climb and walk on, they look like surrealist sculptures. The sea got in on the act too, and worked on the limestone from below, wearing away holes and overhangs; in some cases these are so extreme that a vast mass of rock sits on the water supported only by a narrow stalk.

Miniloc Island seen from Pangulasian Island

Limestone spire on the beach at Entalula Island

Cliffs overlooking the beach at Entalula Island

Cadlao Lagoon, Cadlao Island

Some of the lagoons and beaches of El Nido are so well hidden that you can get to them only by swimming or wading through a narrow opening in a high rock wall, although other gaps are wide enough for a sea kayak. But once you reach them, you will find they are like a catalogue of everything you could ever want in a tropical island. Pristine crescent-shaped beaches are sheltered at each end by limestone stacks standing silent sentinel; vegetation hangs lushly from sheer walls that continue vertiginously down through deep lagoons; and you may even see sea turtles swimming through crystal-clear water.

It takes only a few seconds to swim or snorkel from the rolling swell of Bacuit Bay through a narrow gap that leads on to Secret Beach on Matinloc Island, but what you see is astounding. Suddenly you are in shallow aquamarine water facing a white beach. Below you the sand is dappled with coral and above you near-vertical limestone walls ring the lagoon. The virgin tropical jungle that tumbles down the cliff behind the beach makes the sand appear even more crisp.

There are 25 mammal species on Palawan, 11 of which are endemic. The remainder are species normally found on the island of Borneo, supporting the theory that Palawan was connected to Borneo until about 160,000 years ago. There is evidence that Palawan's birdlife shares this historical link, with 35 per cent of bird species normally found on Borneo. Anthropologists believe that cave dwellers made a similar journey – fragments of bone from so-called Tabon Man, dating back 50,000 years, have been found here.

The Spanish came to Palawan in the 16th century, and according to a popular theory one of their legacies is the name itself, which is derived from what they called the territory: Paragua, meaning umbrella, because the main island looks like a closed umbrella. However, some people believe the origins are Chinese – there is

Cadlao Lagoon, Cadlao Island

Exploring Cadlao Lagoon, Cadlao Island

El Nido Resort, Miniloc Island

evidence of a great deal of trade with China before the Spaniards arrived – and others that they are Indian. The influence of Spain can still be seen all over the island in a number of colonial buildings, including forts and lighthouses. The Spanish left in 1898, following a revolution, and the United States of America stepped in and set up a civil government, which later created the province of Palawan.

Apart from El Nido, Palawan is famous for the underground St Paul River closer to the capital, Puerto Princesa, in the Puerto Princesa Subterranean River National Park. This 8-km cave system has crystal stalactites and stalagmites and it is possible to canoe its entire length.

ⓘ ....

Many international airlines fly to Manila in the Philippines (including Etihad, who fly there direct from London), and from there it is a 90-minute flight to Puerto Princesa, Palawan's capital, with the Island Transvoyager. El Nido's two resorts, Lagen and Maniloc, are 45 minutes by boat from Puerto Princesa.

View of Bacuit Bay from Vigan Island

Small Lagoon, Miniloc Island

Waterfront of the Stonetown

Lamu
Kenya

Lamu is the oldest and best preserved Swahili settlement on the Kenyan coast. Much like its larger and better known cousin Zanzibar, to the south, its African culture blends atmospherically with the influence of Islam, which was introduced to East Africa by traders from Oman in the late 7th century.

It is one of a group of four isolated islands called the Lamu Archipelago, and travelling there is an adventure in itself. It has no international airport and propeller-driven planes from Nairobi land on nearby Manda Island where the grass airstrip is basic to say the least. The transfer to Lamu is by motorized dhow.

Ancient Arabic-style boats, dhows are the traditional form of water transport and are made from wooden planks held together by wooden pegs. Most aren't motorized and you still see them under sail, coming through the channel between Lamu and the outlying islands in the group – often at a crazy angle balanced by a boat captain standing on the end of a long plank.

Life on the island is centred around the waterfront of the old town (also called Lamu) and this is lined with dozens of dhows. Although some are run for tourists by 'boat boys', often with Rasta hairstyles and a Bob Marley fixation, most are used for fishing or for ferrying goods up and down the coast, and reputedly to and from the Middle East. Dhows are also used to carry building materials, including coral blocks – a feature of the island's architecture – from a quarry a few kilometres away.

Once the the boats are unloaded donkey transport takes over. If dhows signify Lamu on the water, it is this humble animal that rules on the land! Sometimes ridden by drovers whose legs seem to reach the ground, the donkeys generally carry materials for the many buildings that are being renovated.

Lamu town was founded in the 14th century, but most of its historical buildings date to the 18th century. They are a mix of

Lamu airport

Waterfront of the Stonetown

Mosque, Shela village

Lamu waterfront

Dhow on the waterfront

Vegetable market

traditional courtyarded houses made from the ubiquitous coral blocks, buildings that have been refurbished as guest houses and others in a state of disrepair. Looming over them and set at the far side of a shady square is the old fort.

In the narrow network of alleys around the fort, life is lived much as it has been for generations. Although most of the small shop-houses sell all manner of goods for the locals there are a number that are aimed just at the emerging tourist trade. It is worth wandering around here for a few hours – you will get hopelessly lost, but that is all part of the fun. It is a good idea to employ a local guide at least once. It is easy to find one – indeed, hard not to – and he will show you places you have little chance of discovering on your own.

The Muslim influence is strong. There are over 20 mosques on the island, including the Riyadha Mosque, which hosts the Maulidi Festival in the third month of the Muslim calendar each year to celebrate the Prophet's birthday. The Pwani Mosque, the oldest on Lamu, dates back to 1370. In the town you will see the *bulbuls* – women dressed head to toe in black, despite the heat.

For a change from Lamu, you can spend a day on Manda Island visiting the ruins of Takwa, a town that was destroyed in the 17th century, and picnic on the adjacent beach. The ruins are atmospherically set among bloated baobab trees, and were built of the same coral as much of Lamu town. There are persistent rumours, though, that lions have swum to the outlying islands from the mainland – a sobering thought if you run aground in the mangroves as a result of mistiming the boat trip, and have to wade through the mud to get to the ruins.

For all its isolation the island has been discovered by the jet set and several celebrities, who have bought a number of houses in the old town. Although renovating them gives work to local people they are worried that the nature of Lamu will change for ever.

Takwa ruins on Manda Island

Donkeys carrying rocks

ⓘ ··

Ethiopian Airways, one of the most modern and well-run airlines in Africa, flies to Nairobi with a stopover in Addis Ababa, which means you can combine a visit to Lamu with one to the Islands of Lake Tana (see page 36). From Nairobi, Air Kenya flies daily to Lamu. The classic place to stay is Peponi's, a short boat ride away from the old town in the quiet village of Shela at the head of an expansive sandy beach.

Big Island
Hawaii, USA

Big Island is arguably the most spectacular of the Hawaiian islands. It is home to the highest mountain on earth – measured from the seabed Mauna Kea reaches 9700 metres – and also to Mauna Loa, which is 30 metres lower than Kea but together with its underwater mass it is taken to be the largest object on the planet. The island's distance from Honolulu's international airport means it is also refreshingly free from the excesses of tourism seen on the more popular islands.

Keokea Beach Park, Big Island

Akaka Falls, Big Island

Sometimes known as Hawaii, Big Island is the youngest of the islands in the Hawaiian group – so young that some areas are still volcanically active. It is made up of five overlapping shield volcanoes, including Kohala, now extinct, and Mauna Loa and Mauna Kea. Kilauea, on the flank of Mauna Loa, has been erupting continuously since 1983.

A unique combination of north-east trade winds and elevation means the summit of Mauna Kea provides some of the world's

clearest conditions for astronomy. There are 13 observatories here with telescopes that are among the most sophisticated ever made. It is possible to drive to them, but a four-wheel-drive vehicle is essential as the road is all but impassable without one, and car rental companies specifically exclude it if you hire any other kind of car. You should also be aware that Mauna Kea's tremendous altitude – the volcano rises to 4205 metres above sea level – could impair your driving skills and even lead to acute mountain sickness (AMS).

North Shore, Oahu

Waipi'o Valley, Big Island

Big Island has 11 of the earth's 14 climate zones, and as you drive around it you will see many of the resulting terrains, from beaches and lush vegetation to lava deserts and even alpine tundra.

Hilo, the main city on the wetter eastern side of the island, is a good place to start your tour. As you drive north the road loops through tropical rainforest towards the Akaka Falls, where a plume of water falls 128 metres into a round pool. At the northern end of this coast is the Waipi'o Valley overlook, where steep cliffs channel a flat valley floor out to the open sea. This valley was the birthplace of the legendary chief Kamehameha the Great, who unified the Hawaiian islands in 1810. From the overlook you can also see the cliffs of the other valleys that scour Kohala, all inaccessible by land.

Sunset, South Kohala coast, Big Island

Inland from Kohala is the cattle-rearing centre of Waimea, home to Parker Ranch, one of the largest ranches in the world. Incongruously,

Sunset Beach, Oahu

cows graze on open pasture in the foreground against views up to the summit of Mauna Kea and its observatories.

The terrain on the drier Kona coast to the west is the result of recent lava flows and there is little vegetation. Tourism is centred on this part of the island, around the town of Kailua and luxury resorts to the north. To the south, the slopes of Mauna Loa are a

Waimea, Big Island

Kahua Ranch, Big Island

Kilauea Iki Crater, Big Island

prime coffee-growing area that produces Kona, one of the world's best-known coffees. Further south you will come to Kealakekua Bay, where an obelisk marks the spot where Captain Cook was killed in 1779 during a fight with local people.

The Hawaii Volcanoes National Park in the south of the island, geologically the youngest part of the all the Hawaiian islands, is home to two of the most active volcanoes in the world. At its heart lies the 1200-metre Kilauea, on the side of Mauna Loa, which looms almost 3000 metres above the rim of the smaller volcano's caldera. The heights of these mountains are almost inconceivable.

A drive around the Kilauea caldera takes you through one of the most desolate landscapes on earth – one that looks more like a moonscape, except for signs that indicate the date of each lava flow. The road circles the sheer cliffs that plunge to the continuously steaming floor of the crater.

The island of Oahu, with its capital Honolulu, is not as striking as Big Island, but all the interisland flights pass through it and it is worth taking the time to stop off and head for the North Shore – the surfing Mecca of the world. Winter storms in the northern Pacific create swells that are amplified by the shoreline and regularly result in waves that reach a height of 6 or 9 metres – and some that soar to 15 metres.

Legendary surf spots here include Sunset Point, the Bonzai Pipeline and Waimea Bay. Sunset is a good time to watch as the surf dudes line up waiting for the big wave, then take their lives in their hands and ride the violent wall of water.

(i) ....

Honolulu International Airport on the island of Oahu is a major airline hub in the Pacific and a number of flights to the USA, Asia and the Pacific Rim land there. Aloha Airlines and Hawaiian Airlines fly from Honolulu to Big Island.

Slopes of Mauna Kea, Big Island

A telescope in the Submillimeter Array, summit of Mauna Kea, Big Island

Britain and France fought over St Lucia 14 times in the 150 years leading to 1814 and, politics apart, it is easy to see why: the landscape is stunning. More mountainous than the other Caribbean islands, St Lucia's highest point is 950-metre Mount Gimie, but its most notable feature is the twin Gros and Petit Pitons, which peak at 798 metres.

The Pitons and Soufrière

Most tourism is centred on Rodney Bay in the north-west, but it is worth hiring a car to explore the rest of St Lucia. The road around the island follows the coast for much of the way, but also takes you inland where it winds up into the lower slopes of the rugged and mountainous interior, much of which is covered with impenetrable, lush rainforest.

There is a very traditional side to St Lucia. Gros Islet, just to the north of Rodney Bay, is an unspoilt Caribbean village and Friday nights here, when the weekly 'jump-up' street party happens, are unmissable. Stalls sell jerk chicken, there are loud sound systems, and tourists and locals alike dance and drink cold Piton beer and large slugs of the local rum.

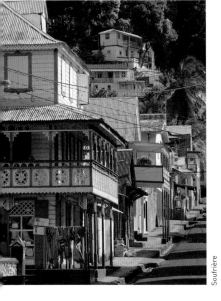
Soufrière

Marigot Bay

Fishing boat in the harbour at Soufrière

North of Gros Islet is the Pigeon Island National Historic Park. This peninsula was the home of the pirate Francois 'Wooden Leg' Le Clerc in the mid-16th century. It was also the site of the British stronghold from which Admiral Rodney launched a 36-year struggle against the French; in 1814, this brought France's possession of St Lucia to an end. It is still possible to stand on the ruins of the fortifications and enjoy commanding views as far as the island of Martinique.

Soufrière is the oldest town on St Lucia and was its capital when it was ruled by France. The main square has a typical French colonial feel, with many surviving buildings. Luxury hotels in the nearby hills and on the coast have stunning views of the Pitons. Synonymous with the island, these are eroded lava domes and, together with the ridge between them, they support eight rare plant species and five endemic bird species. An impressive way to appreciate the twin peaks is to take a scenic helicopter ride; this skirts around them and skims low over the verdant rainforest that coats their lower slopes.

There are a number of reserves where you can visit pristine rainforests. The 7690-ha Central Rainforest Reserve lies at the foot of

Anse La Raye

The Pitons

Balenbouche Estate

Mount Gimie and encompasses rainforest, cloud forest and woodland – all of which can be seen as you climb the mountain's slopes. The Enbas Saut trail leads to two waterfalls on the Troumassee River.

Further south, on St Lucia's south-west coast, is the tranquil Balenbouche Estate with its historic plantation house. The rusted remains of an 18th-century sugar mill have been overgrown by the trunks and trailing roots of massive ficus trees, which seem to grow out of its very walls. Ficus are also known as strangler figs and they really do look as if they are strangling this crumbling ruin.

Although tourism is important to St Lucia, the island's main foreign currency earner is bananas, which are grown in plantations in its interior. They play a major role in the cuisine of the island, so you may find yourself dining out on banana salad, or boiled green bananas and saltfish, while watching the sun set behind the mighty Pitons.

ⓘ ....

The dry season, from January to April, is the best time to visit St Lucia. There are direct flights from Europe and the USA to Hewanorra International Airport, about a 90-minute drive from Rodney Bay. Inter-Caribbean flights use the George F. L. Charles Airport in Castries.

Unicorn in Rodney Bay

Pigeon Island National Park

Cannon at Pigeon Island National Park

Balenbouche Estate

Mont Saint-Michel
France

The imposing and atmospheric bulk of Mont Saint-Michel makes a memorable impression when you first see it, as part of the Normandy countryside. It dominates, yet integrates with, the surrounding farm land, filling in the gap in an avenue of tall poplar trees, hiding behind rolled bales of straw in a harvested field or looming out of the pink pre-sunrise mist. Even in silhouette it is instantly recognizable.

Mont Saint-Michel is a tidal island in the mouth of the Couesnon River. It is home to a number of old dwellings and is topped with a steepled church and a Benedictine abbey – the archangel Michael reputedly appeared to St Hubert, Bishop of Avranches, in 708 and instructed him to build a monastery here.

Mont Saint-Michel looms over a flock of sheep, seen from Bas-Courtils

Village on the slopes of Mont Saint-Michel

The series of walls and towers on Mont Saint-Michel's perimeter protected it against the English during the Hundred Years War between England and France in the 14th and 15th centuries. They repeatedly attacked the island, but failed to take it because of a combination of these fortifications and its isolation from the mainland. Its defensive nature was turned around during the French Revolution when it was used to imprison high-profile political prisoners. Despite protests by prominent figures, including Victor Hugo, Mont Saint-Michel remained a prison until 1863.

Hugo was a great fan of the island. It was he who coined the phrase 'à la vitesse d'un cheval au gallop' to describe how the tides here race in at the speed of a galloping horse. At more than 14 metres,

they are among the highest in the world. The water rushes in at over 32 km an hour and there is a real risk of being drowned if you venture too far out on the sand that surrounds Mont Saint-Michel at low tide. Be aware, too, that there are also patches of quicksand.

Light floods through one of the fortified gates on the Rue Grande

Village of Mont Saint-Michel

Mont Saint-Michel was originally more of an island than it is today. Historically it was connected to the mainland by a narrow, natural land bridge that was completely under water at high tide. However, the construction of polders to reclaim land for farming, and the permanent causeway that was built towards the end of the 19th century, have led to the bay silting up. There are plans to replace the causeway with a bridge, to reverse this process and make Mont Saint-Michel a true island again.

A giant car park at the island end of the causeway is used as an impromptu campsite for motor homes in summer. The sea of cars, and the crowds on the island and in the small town on the mainland with its hotels and restaurants, can be off-putting. So make sure you visit Mont Saint-Michel at the right time: in the early morning – the narrow alleyways of the village at the foot of the abbey are deserted

Towering apse of the church at Mont Saint-Michel

Looking down the cloisters at the Benedictine abbey

Village of Mont Saint-Michel seen from the causeway

just after sunrise – or in the evening when the crowds have dissipated.

The abbey opens at 9 a.m., long before most visitors arrive, which will allow you to walk around in relative peace – even in the summer months. Look out from the battlements at about 11 o'clock and the people swarming up the causeway will resemble an invading army. This is a good time to head out into the countryside to view Mont Saint-Michel from afar, and appreciate how it fits into its environment.

(i) ···

Mont Saint-Michel is a short drive from Dinard Airport. Aurigny Air flies to Dinard from London via Guernsey, which makes it possible to combine your visit with one to Sark (see page 144). There are a number of hotels on the mainland end of the causeway, but staying on Mont Saint-Michel itself will allow you to appreciate the island's unique atmosphere.

St Aubert's Chapel at the rear of Mont Saint-Michel

Mont Saint-Michel seen from the bay at low tide

Tierra del Fuego
Argentina

Tierra del Fuego, or more correctly Isla Grande de Tierra del Fuego, is the largest island in an archipelago that lies at the foot of South America. The mighty Andes tail out here before slipping into the Atlantic to resurface as the Falklands, South Georgia and the peaks of Antarctica.

Lake Roca, Tierra del Fuego National Park

The island is jointly owned by Argentina and Chile, and Ushuaia on the Argentine side is the most southerly city on the planet. When you arrive here you will feel that you have reached the end of the world. Even the inbound flight is dramatic: the plane drops down from the last peaks of the Andes and flies along the Beagle Channel before touching down. Taking off is even more breathtaking, with a startlingly sharp turn to avoid crossing the border into Chilean airspace.

There appears to have been little or no control over how Ushuaia developed. Houses come in a variety of styles and the city has a real frontier-town feel. Rugged mountains form a spectacular backdrop to this strange place, their size accentuated by the low-rise nature of many of the buildings.

Sunrise, Ushuaia

It is worth hiring a car to explore the island – and a good start to your tour is a drive along the RN3 to the Tierra del Fuego National Park (Parque Nacional Tierra del Fuego), west of Ushuaia. Relatively small, this boasts beech forests, lakes, a picturesque range of mountains and a rocky coastline. The road ends here, just short of the border with Chile.

North from Ushuaia the RN3 climbs quickly to a commanding view across the floor of the Carbajal Valley and the peaks of the Sierra Alvear, then continues up to the dramatic Garibaldi Pass, which offers vistas down to Lago Fagnano. Tolhuin, on the shores of the lake, is the last town before Rio Grande – Tierra del Fuego's second city after Ushuaia – 100 km further north. A number of unpaved roads lead off the RN3 and provide excellent access to the interior of the island. They are secondary routes, designated 'RC' (*ruta complimentaria*).

Estancia Indiana

Beagle Channel, Tierra del Fuego National Park

Grasslands near Estancia Harberton

For a particularly scenic drive take the RC-j off the RN3 and continue some 35 km towards historic Estancia Harberton. The road follows the Lasifashaj River past mountains and old-growth forest and comes out onto the Beagle Channel, with its views across to Puerto Williams in Chile. It then cuts back inland and goes through what can only be described as golden grassland, peppered with trees bowed by constant winds, before reaching Estancia Harberton.

Hills near Lake Yehuin

A missionary family established this historic ranch in 1886 as a place of refuge and peace for the original inhabitants of Tierra del Fuego – whose fires Ferdinand Magellan saw in 1520 when he sailed into what is now known as the Strait of Magellan. However, by the 1930s the population had been decimated by disease and contact with Europeans.

Another spectacularly scenic road, the RC-a, takes you through ranch land to the Atlantic coast at Cabo San Pablo. The wide beach here is home to the *Desdemona*, shipwrecked during a storm in the early 1980s. From the hill above it you will be able to see down to the wild interior of the Mitre Peninsula – the inhospitable end of the continent.

Pia Glacier, Darwin Mountain Range

Beagle Channel, Darwin Mountain Range

A three-day cruise between Ushuaia and the town of Punta Arenas in Chile will provide you with unique views of Tierra del Fuego. The ship passes through spectacular – and otherwise inaccessible – scenery including the peaks and glaciers of the Darwin range. You will even be able to go to the towering faces of some of the glaciers in an inflatable Zodiac. The cruise also goes south to Cape Horn – and round it if the conditions are right. The seas here are renowned for their roughness and sailing past the cape is a never-to-be-forgotten experience.

Ushuaia: the most southerly city on the planet

Penguins National Monument, Magdalena Island, Strait of Magellan

Punta Arenas, like Ushuaia, feels like a frontier town. It is a good point from which to head further north to the spectacular Torres del Paine National Park (Parque Nacional Torres del Paine) and the wilds of Chilean Patagonia.

ⓘ ··

There are flights to Ushuaia from Buenos Aires in Argentina on Aerolineas Argentinas and via Punta Arenas from Santiago in Chile on LAN Chile. Flying between Ushuaia and Punta Arenas is relatively cheap. Cruceros Australis offers one-way cruises between Ushuaia and Punta Arenas in both directions.

Sicily
Italy

The harbour at Cefalù, Palermo province

The largest island in the Mediterranean, Sicily is a smorgasbord of European history. Because of its strategic position it was colonized by the Greeks and Romans, and ruled by Byzantines, Arabs, Normans and, lastly, the Spanish before it became part of Italy in 1860. This led to a mix of bloodlines that has given rise to a culture that is unique to the island: the local people consider themselves Sicilians first and Italians second.

One of the best ways to appreciate all that Sicily has to offer is to hire a car and drive through its many centuries of history. Near the capital Palermo is Monreale, a hill town at the end of the picturesque Conca d'Oro Valley. It has spectacular panoramic views down to Palermo,

Hill town of Castiglione di Sicilia, Catania province

Hills near Segesta, Segesta province

and the sea beyond. The interior of the 12th-century Norman cathedral is covered with mosaics, and when you view them from a distance it is almost impossible to believe they are made just from tiny coloured squares. Their complexity and subtlety are breathtaking, with every major event in the New and Old Testaments depicted in almost overwhelming detail. The cloisters of the adjacent Benedictine abbey have an Arabic flavour blended with Byzantine and Norman styles, indicative of the varied influences on the island.

On the south-east corner of the island, Siracusa is steeped in more than 2500 years of history. The city was founded by Corinthians in 734 BC and later became one of the most important cultural centres of its time –

rivalling Athens in the Greek world. Its medieval heart is on the tiny offshore island of Ortygia.

A number of hill towns to the south-west of Siracusa were rebuilt after a devastating earthquake in 1693, at the height of the baroque period, and the architecture of Noto is some of the finest on the island. Walking through the massive Porto Reale arch and along the Corso Vittorio Emanuele it is impossible not to be humbled by the grandeur of the buildings with their ornate details. Elaborate towers and balconies

Cathedral window, Monreale, Palermo province

Figure of Christ in the central apse of the cathedral, Monreale, Palermo province

The town of Ragusa Ibla, seen from Ragusa

augment façades decorated with gargoyles, cherubs and lions.

The Valley of the Temples in Agrigento, west of Noto, is the largest and best-preserved site of Greek ruins outside Greece. The most striking structure is the Temple of Concord – its walls are largely intact and give a strong impression of the scale and the complexity of the original building. The yellow stone glows golden at the beginning and end of the day.

Further west are the ruins of the Greek city of Selinus. Dating back to the 7th century BC, they cover a much wider area than the

Valley of the Temples and it is possible to visualize the extent of the ancient city. They are on the coast, and a walk amid their Doric columns gives tremendous views out to sea. About 35 km away, isolated in rolling green hills, is the site of the Greek town of Segesta. All that remain of it, perched on a high outcrop, are an amphitheatre and a temple that was started in the 5th century BC but never completed. At times in their histories Selinus and Segesta fought border wars with each other.

Castle above the town of Cáccamo, Palermo province

Garden and quadrangle of the cloisters, Benedictine monastery, Monreale, Palermo province

Detail of a carved pillar in the monastery cloisters, Monreale

Sicily is famous for Mount Etna, at 3323 metres the highest active volcano in Europe. The best way to approach it is through the mountainous interior, past a string of hill towns that are all you would expect them to be on this island: atmospheric jumbles of houses topped by churches and ancient castles. Groves of olive trees stretch across dry valleys. The road weaves around hairpin bends and across viaducts and bridges.

The first views of Etna are stunning: as you look down from the hillside towns broad sweeping valleys lead to forests on the volcano's

South-western edge of the hill town of Gangi, Palermo province

Mount Etna over Maletto town, Catania province

lower slopes and lava flows on its upper ones, topped by snow at its summit. Some of the best vantage points are in the town of Centuripe, south-west of Etna.

It is possible to reach the higher slopes of Etna by road, cable car or even the Circumetnea Railway, which originates at Catania on the east coast, but the most striking views are from afar with the volcano dominating the surrounding countryside.

(i) ....

There are a number of international flights to Palermo. Accommodation can be tricky in the peak summer months, but if you avoid these it shouldn't be necessary to book in advance. The weather is usually warm for much of the year. Hiring a car is the best way to explore the island.

Doric temple at Segesta, Trapani province

Sunrise, Bar Harbor

The character of Mount Desert Island was formed by massive glaciers that gouged out its lakes and valleys and created the rounded profiles of its mountains. These features are on a north–south orientation, mimicking the progress of the encroaching ice sheets two to three millennia ago.

Bar Harbor

Just off the coast of Maine in the north-eastern United States, the island, much of which is gazetted as the Acadia National Park, is vaguely horse-shoe shaped with the two halves encircling Somes Sound – the only fjord on continental USA. The eastern half contains many of the park's most famous features, such as Bar Harbor, Sand Beach and Otter Point, but there are tremendous hikes and a stunning coastline on the less crowded west side.

Human settlement here dates back at least 6000 years and the Wabanaki people called the island Pemetic, or 'sloping land', after the characteristic shape of its mountains. From the early 17th century until

Somesville Bridge, Somesville

the 1760s, when Britain established the first permanent settlement, influence over Mount Desert oscillated between the French and British.

The population increased after the War of American Independence and by the end of the 19th century tourism was a big business. At about this time Mount Desert also became a favoured retreat for many of the great industrialists and business leaders of the era – families with names like Vanderbilt, Rockefeller, Ford and Pulitzer had ostentatious holiday homes here. Many of them were instrumental in the preservation of the island, which eventually led to the establishment of a national park that was later given the name Acadia. John D. Rockefeller, Jr donated much of the land for this, and was responsible for a network of 'broken-stone' carriage roads – roads closed to motorists – many of which are still open to walkers. The Great Depression ushered in the decline of this golden age, and the final nail in its coffin was a fire in 1947 that destroyed many of the estates.

Bar Harbor is the largest of the island's towns and the centre of tourism. The smaller communities are essentially fishing villages and specialize in lobster; they were also famous for cod, but this industry has collapsed in recent years.

The real draw of Mount Desert Island is its stunning natural scenery: the 1.6-km stretch between Sand Beach and Otter Point is arguably the quintessence of the Maine coastline. At sunrise, the rugged granite cliffs and rocky coves glow pink, while a thick bank of fog often rolls in from the sea, diffusing the sunlight. Offshore lobstermen weave from buoy to buoy checking their pots, and flocks of common eider bob in the swell.

A few steps inland on the coastal path the clear, crisp sound of waves breaking on a beach of smooth, round cobblestones is muffled by the stands of pine and spruce trees that grow in thick beds of moss covered with ferns and many types of fungi, including mushrooms.

Brown Mountain Gatehouse, Northeast Harbor

Echo Lake as seen from Beach Cliffs

Lobster traps and buoys, Bernard

The trails and carriage roads that comb the interior of the island in an extensive network often lead up steep valleys to the glacially polished mountains. The views from the flat, rounded summits are spectacular and many subalpine flowers grow in crevices and cracks in the rock.

In autumn the birch and maple trees on the lower slopes are a kaleidoscope of reds and oranges, and at sunset and sunrise their leaves glow as though they are on fire. Above them the tiny silhouettes of lobster boats weave through the bright reflections of the golden sun.

Boats in Bass Harbor, Bernard

Bar Harbor

Autumn colour, Acadia National Park

(i) --

Colgan Air flies from Boston international airport to Hancock County–Bar Harbor airport, 1.6 km from Mount Desert Island and 13 km from Bar Harbor. The free Island Explorer shuttle-bus network runs from mid-June to Columbus Day holiday (the second weekend in October) and covers most towns, campgrounds, park locations and trail heads. It is difficult to predict the exact timing of the autumn-leaf colour, but it generally peaks between mid-September and mid-October. Carriage roads are still closed to motorized vehicles. Balance Rock Inn has stunning views over the coast.

Svalbard
Norway

Mountains in Hornsund fjord, Spitsbergen

The Svalbard archipelago lies between mainland Norway and the North Pole and is perfect for exploring the Arctic, one of the most remote and hostile regions in the world. Here, at the very top of the northern hemisphere, the sea is permanently covered with ice – even during summer when the sun never sinks below the horizon. The winter months see the seemingly endless darkness of the polar night.

The snowy wastes of the Arctic don't have the same amount of wildlife as the Antarctic with its vast hordes of penguins and almost tame nesting birds, but the region does have an edge on its southern counterpart. It is the realm of one of the world's top predators: the fearsome polar bear.

MS *Explorer* ploughing through the pack ice

MS *Explorer*

In the summer months the MS *Explorer* cruises the waters around Svalbard in search of these and other Arctic animals. Trips start and finish at Longyearbyen, the biggest settlement on Spitsbergen, the main island of the archipelago. These are expedition voyages, and the schedule is fixed only by the state of the pack ice and the quest for mammal and birdlife.

A strange but engaging place in the most tremendous setting, Longyearbyen is so far north there is no daylight during winter; a number of its houses are painted jaunty, bright colours, no doubt to outweigh the effect of months of the polar night.

Spitsbergen is so isolated that the Norwegian government is seeking to establish a doomsday seed bank here, in order to

Mountains in Hornsund fjord, Spitsbergen

Three polar bear cubs on the pack ice

preserve biodiversity should there ever be a man-made or ecological disaster.

There are a number of totally deserted islands in the Svalbard archipelago, with plunging glaciers and snow-covered mountains. The area is pristine and you'll feel as though you really are heading into the wilds. The polar bears aren't corralled into a shrinking area by melting ice, as they are at Churchill in Alaska, where they are so

Freemansundet

Svalbard reindeer

easy to spot they might as well be in a zoo. Here they roam over kilometres of ever-changing pack ice, hunting for seals.

Finding the bears is a challenge and seeing them is a unique thrill. We came across a mother with two almost fully grown cubs who, rather than move away from our looming red ship, stayed on their ice floe and drifted closer and closer to the *Explorer* until they were sniffing inquisitively just by its hull. Completely unperturbed by our presence, they vocalized their curiosity with prolonged roars that echoed hauntingly across the barren emptiness.

On another occasion we saw a mother with not two but three young cubs – almost unheard of in this part of the Arctic. As they moved away they went from one ice floe to another, swimming

Walrus on the pack ice

between them and then rolling on the ice to expel the water from their shaggy coats. At one stage a cub hitched a ride on its mother's broad back as she swam through the freezing sea.

The expedition includes a number of excursions by inflatable Zodiac boats, amid ice floes and even right up to the sheer faces of giant glaciers. Some involve landing on an island and trekking across frozen wasteland, always with armed guards as escorts – a reminder that human beings are uninvited guests in polar bear territory.

On these shore visits you may occasionally come across trappers' huts. Long deserted, these are typically basic, wooden constructions strengthened against attack from polar bears. Many of the trappers stayed on the islands, alone in the total darkness, throughout the winter months when animals such as the arctic fox sported their thicker, white winter fur.

Polar bears are not the only great mammals in this region. The

Colourful houses at Longyearbyen

Trappers hut at Kapp Toscana

mighty walrus lounges on the pack ice and deserted beaches until it overheats and has to drag its vast body back to the cooling waters. Squabbles break out between these garrulous and argumentative animals, and great tusks are brandished in threatening poses before laziness beats temper and everything settles down again.

Until it was granted to Norway in 1920, Svalbard was an open region exploited by a number of nations. The Svalbard Treaty gives anyone who can find employment the right to settle here and make use of the islands' natural resources. There is still a Russian presence on Spitsbergen – Barentsberg is a settlement of around nine hundred people. Pyramiden, a deserted mining town on the same island, gives an eerie idea of what life was like in the Soviet era – it comes complete with a bust of Lenin and even a remote hut made entirely of bottles that once contained alcohol.

Lenin bust at deserted Russian settlement of Pyramiden

ⓘ ··

The MS *Explorer* is owned and run by GAP Adventures, who organize a number of cruises in summer. Their Realm of the Polar Bear trip starts and finishes in Longyearbyen on the island of Spitsbergen and explores the more remote regions of the archipelago, depending on the state of the pack ice. Scandinavian Airlines (SAS) fly to Oslo and then on to Tromso and Longyearbyen.

Heleysundet, between Spitsbergen and Barentsoya

Statue of Columbus below Government House, Nassau

Of all of the 660 uninhabited and 40 inhabited islands that make up the Bahamas, arguably the best beaches can be found on charmingly rustic Eleuthera, a narrow strip of land nestling between the Atlantic and the clear warm waters of the Caribbean. At 160 km long, but no wider than 3 km at any point, it certainly possesses enough coastline to choose from, and without the rampant overdevelopment of many islands in the Caribbean.

Bahamas is a derivation of the Spanish *baja mar* (shallow sea), and the combination of this lack of depth and brilliant white sand means Eleuthera is surrounded by some of the most deeply turquoise water you will see anywhere in the world. Its towns are rather run-down, but in an atmospheric way, with whitewashed churches vying for space on the seafronts with wooden-decked beach bars and shops selling colourful local paintings.

Beach at Tippy's Restaurant, near North Palmetto Point

Dunmore Town, Harbour Island

On the north of the island the Bight of Eleuthera takes such a chunk out of the Caribbean coast that its waters almost meet the great surging waves of the Atlantic. These try to force their way through a gap no wider than 25 metres, to reach the tranquil waters on the other side. The Queen's Highway, which runs the length of the island, spans this narrow with the well-worn Glass Window Bridge, which has been repaired a number of times after being washed away by countless Atlantic waves.

Just a few kilometres from Eleuthera lies the smaller and more upmarket Harbour Island. The main attraction here is the famous Pink Sand Beach, but you have to be there at sunrise or sunset to really appreciate the colour. The beach runs almost the entire Atlantic side of the island, and along the dunes behind it is some of the most expensive real estate in the Bahamas.

Although Harbour Island has seen an influx of money over the last 10 years it has retained its charm. Dunmore Town, with its neat clapboard houses fringed with white picket fences, is its only settlement. It still has a fishing community, and because there are only a few cars most of the tourist transport is on golf carts.

If you are travelling between any of the Bahamas' main islands you will generally have to spend some time in the capital, Nassau, on New Providence. This is not a hardship, however – Nassau is one of the most genial cities in the world.

From the steps of the statue of Columbus, below pastel-pink Government House, you can look out across the buildings and harbour and see a slice of the way Nassau has changed over the years. Painted wooden houses give way to the Georgian Parliament Square with its statue of Queen Victoria. Gliding in the background are the new colonizers of the Bahamas: massive cruise ships, pulling into Prince George Wharf.

Bay Street, downtown Nassau

St Columba's Episcopal Church, Tarpon Bay, Eleuthera

Nassau is a relaxing place for a stroll, or you can take one of its horse-drawn carriages. This is one of the best ways to see the city's famous traffic cops, who stand at intersections in crisp white tunics and signal to the passing vehicles with insistent and precise movements of their white-gloved hands.

ⓘ ··

Virgin Atlantic flies from London to Nassau. From here, Bahamasair flies to Governor's Harbour on Eleuthera, and Bahamas Fast Ferry runs a service to Dunmore Town on Harbour Island. Bizarrely, to fly to other Caribbean countries, such as St Lucia (see page 168), you have to connect through Miami. The Coral Sands Hotel is one of just two hotels on Pink Sand Beach and has public beach access.

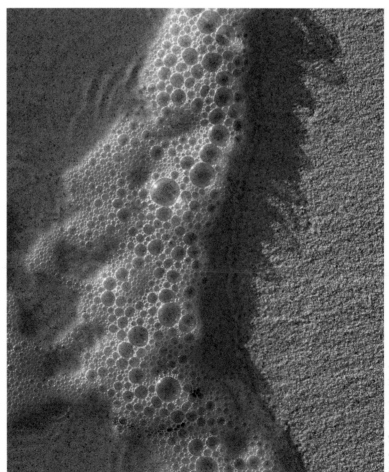

Pink Sand Beach, Harbour Island

Lone Tree on Harbour Island

Ibiza and Formentera
Spain

Fortifications of Dalt Vila, Evissa (Ibiza town)

Most people have an opinion about the Balearic island of Ibiza: clubbing Mecca, haven for raucous holidaymakers or a budget destination for families. While all this may be true, it is also beautiful and undiscovered, steeped in history, peppered with ancient terraced farmland and surrounded by a coastline that in parts is as rugged, unspoilt and deserted as anywhere in the Mediterranean.

Finding this aspect of Ibiza (Evissa) takes a little effort – and certainly a hire car. Most of the unspoilt areas are off the beaten track and not easily reachable by public transport – which is what keeps the

summer hordes away. If you are feeling fit, a network of cycle paths links the best parts of the island.

The areas in the north-west and south of Ibiza are some of the least developed, with rutted roads that pass through quiet farm land. Here goats feed under carob trees that are planted in russet-red soil. A characteristic of the countryside is the terraces on the hillsides, reinforced by yellow dry-stone walls made from Ibizan limestone – many of the buildings on the island, and even the cliffs, are the same colour.

There are a number of tranquil villages, such as Sant Miquel and Sant Augusti, which are usually built around an oversized, whitewashed

The rugged south coast of Ibiza

church. Many of the churches are fortified, reflecting Ibiza's turbulent history.

The island was first inhabited in Neolithic times, and later settled by Phoenicians and Carthaginians, both of whom regarded it as strategically important in the Mediterranean. The Carthaginians were the first to fortify the Dalt Vila, the historic old quarter of Ibiza town. They were also the first to recognize the spirituality of the island; they

Village of Sant Miquel

Església de Sant Miquel

Traditional dry-stone building in South Ibiza

believed that anyone who was buried here had a direct path to the afterlife, and imported thousands of bodies.

The Romans came next, bringing development and prosperity in a rule that lasted some five hundred years. They were followed by Vandals, Byzantines, Viking and Norman raiders, and, finally, Moors from North Africa who conquered the island in 902. The Catalans, and later the Castilian Spanish, recaptured it over three hundred years later but it was plagued by pirates, which resulted in a network of defensive towers being built around the coast.

Windmill on Formentera

Old building in Dalt Vila

Façade in Evissa (Ibiza town)

Although it is a well-visited island, Ibiza has several unspoilt beaches, some of which you can have to yourself, and stretches of rugged coastline with flat slabs of rock – perfect for sunbathing. In the north-west there are isolated coves, including Cala d'Aubarca. In the south the exposed peninsula to the west of Sant Antonio, one of the island's tourist centres, encompasses some of the most iridescent turquoise water you'll find outside the tropics. Many of the typical fishermen's huts that line its beaches and coves date back years, and some are even two storeys high, with living quarters above a boathouse.

Sailboat passing Es Vedra

Sunrise at Evissa (Ibiza town)

If you want to be sure of finding a deserted beach head to Formentera, and its long golden sands offset by clear turquoise waters. A small island less than an hour's ferry ride away, it is flatter, drier and quieter than Ibiza.

Although the town of Sant Antonio has been swamped by mass tourism and has little to offer, Ibiza town is elegant and stylish, and the Dalt Vila is surrounded by what are said to be some of Europe's best-preserved fortifications – dating back over four hundred years they are up to 25 metres high. Local people still live in its cobbled alleys. Lively with restaurants and bars at night, these are more atmospheric in the early morning when tourist businesses are closed and the streets start waking up.

Es Vedra, a 378-metre limestone rock, lies just off the south coast. It was sacred to the Carthaginians and is reputed to be the island of the sirens in Homer's *Odyssey*. The hippies who came here in the 1960s claimed it was the most magnetic place on Ibiza, and the rock was the location of many UFO sightings. Whatever the truth, the defensive tower on a high cliff opposite Es Vedra is one of the most evocative places from which to watch as the sun sinks slowly into the sea.

At sunset, as at sunrise, the island comes alive – everything seems to glow red in the Mediterranean sun – and the pleasant cove of Benirràs is another great place to be at this time of the day.

ⓘ ···

July and August are crowded and finding accommodation can be difficult. June and September are easier. Spring is a beautiful time to be on Ibiza. If you are in a group, it can be very cost-effective to hire a villa. Otherwise there are a number of *agroturismo* – country farmhouses that have been converted into boutique hotels – such as the luxurious Can Lluc, in a quiet valley in the middle of the island.

Rugged coastline near Portitxol in the north-west

Sunset at Benirràs Beach

Kubu Island
Botswana

Flowering succulent plant on the top of Kubu Island with the pans behind

The granite island of Kubu lies in the Sowa salt pan in the great Makgadikgadi Pans National Park in Botswana. The area is the site of one of the ancient world's largest lakes, which silted up thousands of years ago, and is a vast and empty void the size of Switzerland. As you travel endlessly across its bleak emptiness you will realize it is literally an eternity of nothingness. The cracked white crust of the pan merges into a blinding, featureless glare and mirages shimmer, looking for all the world like cool, thirst-quenching lakes.

The sight of Kubu on the horizon is a welcome relief. More correctly known as Lekhubu (ridge, in the local Setswana language), it is a gnarled granite outcrop topped with ancient baobab trees. Some of these are monsters, older than Christianity – they have been growing here for over two thousand years. Their gnarled and wizened branches reach out, offering some shade in this parched region.

Kubu is particularly atmospheric at sunrise and sunset, when the granite is tinged pink and the baobabs glow red as if they are about to burst

Looking towards the granite bulk of Kubu Island

into flames. The horizon is so flat that the sun sinks into, or rises from, it as if it were water, spilling a burning mirage like liquid over the land.

The island is also striking by moonlight, as the baobabs cast eerie, long, spindly shadows and all colour seems to drop away in the flat, pale light. But be careful walking around at night. Although there is no wildlife most of the time, animals such as hippo and antelope, drawn by mirages, occasionally wander over the pan. Sometimes they find their way back to its edge; at other times they wander off to die in the great emptiness, leaving tracks that tail off into the distance.

In places the surface of Kubu is stained white with guano, testament to the ancient bird population that used to live here before the lake dried up. Fossils of marine creatures can be found, and pebbles rounded by the action of waves. But as you stand on the island's summit, looking out into the vast nothingness, it is inconceivable to imagine Kubu surrounded by water. The horizon is so empty, and so silent. All you are able to hear is the wind in your ears and the sound of your breathing. Yet, if the annual rains are heavy the pan will flood,

Giant baobab trees at Kubu Island

albeit with just a few short-lasting centimetres of water, and thousands of flamingos will congregate near the island to nest.

The rainy season generally lasts from November to February, although it is unpredictable and can last longer. During this time it is impossible to get to Kubu. The surface of the pan is treacherous and even a four-wheel drive will be bogged down in minutes. However, a couple of months after the rains, when the island can be reached again, there should still be flamingos only a few kilometres away.

Ostrich silhouetted by the setting sun near Jack's Camp

Kubu has been sacred to many peoples over the centuries. The remnants of Stone Age cutting tools and pottery shards dating back two millennia have been found here, and there is also a stone enclosure, believed to date back to the Great Zimbabwe dynasty of the 14th to 15th centuries, that was used as a circumcision area. Cairns left by initiates can still be seen. Later Kubu became a sacred site for the San. Although it was too late in their history for them to leave rock paintings, they left beads made of ostrich shell as offerings between the great boulders that once formed a cave on the highest end of the island.

Over the centuries the San have been persecuted throughout

southern Africa and there are precious few who still live in the traditional way. Many were killed by early European settlers and still more have recently been moved from their land to resettlement camps where they fall prey to alcoholism and drug abuse. However, there are people with San blood living in the area adjacent to the pan, and they still sometimes travel to Kubu to leave offerings.

The island is an intensely spiritual place, and the absence of the usual distractions seems to make this more exaggerated and

Flamingos in flight near Kubu Island

Clouds lit by the setting sun over Kubu Island

Baobab trees silhouetted by the sunset

poignant. Silence in the modern world is difficult to come by – and here it is absolute. The link with history is palpable, and it does not take much to imagine spirits moving through the great baobabs that have stood sentinel on Kubu for so many generations.

ⓘ ··

Uncharted Africa runs Quad Bike Safaris to Kubu Island from its legendary Jack's Camp. Camping facilities at Kubu are very basic, but the company provides all the comforts of home including comfortable sleeping bags, hot showers and an alfresco bar. The food is incredible. Don't miss the chance of a couple of nights at Jack's Camp afterwards.

Looking from the island of Bryher across to Cromwell's Castle, Tresco

Phone box on St Martin's

The Isles of Scilly are a cluster of more than a hundred islands off the tip of Cornwall, some 45 km from Land's End. They are part of Britain, but their sub-tropical climate means they feel anything but British.

The islands were first settled 4000 years ago and this has left them with a legacy of the largest concentration of archaeological sites in the United Kingdom. Over a thousand have been recorded, including standing stones (menhirs), a Bronze Age village and even Roman remains. Neolithic burial chambers are seemingly everywhere.

Many of the Scillies were joined when sea levels were lower, and some people believe they could have formed one large island. Certainly, the remains of field walls are uncovered at low tide.

There are five inhabited islands – Tresco, Bryher, St Martin's, St Mary's and St Agnes – each of which has its own character. On all of them, however, things move slowly. They are linked by small boats and you will find honesty boxes alongside local produce like vegetables, eggs – and even live crabs – or souvenirs such as glass buoys.

Deserted beach on Tresco

Abbey Gardens, Tresco

St Martin's from the air

The island of Tresco is best known for its Abbey Gardens – a vast collection of exotic and subtropical plants. Many were brought back from far-flung parts of the world by sailors and are testament to the seafaring traditions of the island. The imported seeds and plants thrive in its balmy climate. Things to see in the gardens include a sea of giant tree ferns from New Zealand, a wall of cactus and aloes from southern Africa – and the overgrown ruins of the former abbey. Even if you are not particularly interested in plants it is possible to spend hours walking around this tranquil spot. Adjacent to the gardens is the Valhalla Shipwreck Museum – a collection of figureheads from ships that were wrecked on the treacherous rocks off the Scilly islands.

A narrow channel between Tresco and the east coast of Bryher is guarded by a fortified tower, known as Cromwell's Tower. The Isles of Scilly were a royalist stronghold during the Civil War, but were captured by Parliamentarians in 1651. The islands' connections with royalty are still strong. They are owned by the British Crown, and currently administered by the Duchy of Cornwall.

Tresco harbour from Bryher

Looking across to the Round Island Lighthouse from Tresco

Cormorants on the Western Rocks

Tresco harbour

Bryher couldn't be more different from Tresco. Far less busy, it is shaped by the Atlantic gales that lash it in autumn and winter. The spectacular and rugged coastline of Hell Bay and the Shipman Head Downs in the north-west of the island are covered in tough heather and bracken.

St Martin's is noted for its long and all but deserted beaches. There are some stunning walks around this island, and there is even a vineyard that produces local wine.

St Mary's is the largest of the Isles of Scilly but even so it has only 14 km of roads. Although it is the busiest island, the pace is still soporific. Its harbour is dominated by the Garrison and its 16th-century Star Castle. Now a hotel, this was originally built at the time of the Spanish Armada and is shaped like an eight-pointed star.

A sunset cruise from St Mary's to tiny St Agnes will drop you conveniently at the Turk's Head, the most south-westerly pub in Britain. There are also tours around the uninhabited eastern isles where you will have a good chance of seeing basking seals.

(i) ··

British International Helicopters has a shuttle service between St Mary's and Tresco, and Penzance on the mainland. There is also a ferry. There are scheduled boat services between the five inhabited islands (Tresco, Bryher, St Mary's, St Agnes and St Martin's). From these you can take short boat rides to many of the uninhabited ones. Tresco Estates run two spectacular hotels: the Island Hotel on Tresco and Hell Bay Hotel on Bryher. Both of these boast excellent cuisine.

Abbey Gardens, Tresco

Phang Nga Bay Islands
Thailand

The Thai islands are a byword for golden beaches backed by lightly swaying palm trees and gently lapped by crystal-clear water, often with spectacular coral reefs offshore. And some of the most unspoilt are in Phang Nga Bay. Off the west coast of Thailand, in the Andaman Sea, the bay also has what is arguably the country's most breathtaking scenery: the towering limestone outcrops that rise from its waters have been eroded by rainfall until they resemble dripping, melted wax rather than solid rock.

The Paradise Beach Boutique Resort on Ko Yao Noi, one of the largest of the islands, is an ideal base for exploring some of Phang Nga's

Looking out to Phang Nga Bay from Ko Yao Noi

Paradise Beach, Ko Yao Noi

Paradise Beach, Ko Yao Noi

Islands in Phang Nga Bay

myriad secluded coves and deserted beaches. Not only is it set on a perfect stretch of sand atmospherically dominated on one side by a giant limestone crag, but it looks over a vista studded with the bay's signature rock formations.

The waters in Phang Nga are tidal and the sea goes out a long way. This affects the islands drastically: when the tide is low they can be linked by great spits of sand that allow you to walk between them; when it is high, many beaches are all but submerged. On many of the islands there are *hongs* (lagoons) or secluded coves almost completely enclosed by high craggy cliffs.

It is worth hiring a sea kayak to explore Ko Yao Noi and some of its nearby islands. Ko Kudu Yai has a classic cove that is accessed through a narrow opening in its cliff walls. These have created a lagoon that is home to colourful living corals. At high tide the beach becomes a narrow strip of sand.

Ko Roi's amazing interior is accessible only at low tide, when the retreating waters uncover a tiny opening in the cliffs behind the beach. Duck through this gap and you are in a hidden world of mangroves and dense vegetation – open to the sky, but surrounded by tall walls of rock – that is partially flooded when the tide is high.

Further away, when the tidal waters flood Ko Hong's lagoon it becomes large enough for boats to cruise around the interior of the island. Closer to the mainland, Ko Gai has been eroded to such an extent that it resembles a chicken crouching in the water.

Phang Nga is not just the preserve of tourists. Fishermen and even sea gypsies live in the area, and you will often see their characteristic high-prowed craft plying the waters between the islands – coloured scarves are tied around the prows to ward off evil spirits. Many of the boats can be hired by the day, which will enable you to visit the more hidden corners of the bay.

Beach near Krabi on the mainland

Lagoon at Ko Hong

Islands in Phang Nga Bay

There are also a number of caves where local people farm tiny birds' nests, which are exported to China as ingredients for the famous soup. These valuable sites are guarded by small encampments, which you are welcome to visit in return for a small donation.

Once your touring is done, head to the top of Ko Yao Noi for some of the best views of Phang Nga Bay. They are especially evocative at sunset when the sun has disappeared and the sky turns pink, then purple, before darkness falls.

ⓘ ....

Etihad Airways flies to Bangkok in Thailand from a number of countries worldwide. Audley Travel tailor-make itineraries for Phang Nga Bay and the surrounding areas.

Fishing boat off Ko Roi

Narrow spit of beach in Phang Nga Bay

Boat moored by limestone island

Socotra
Yemen

Crabs on the beach at Erher

Socotra lies in the Arabian Sea, off the Horn of Africa, and is the main island in a small archipelago of the same name. Unlike most other remote islands, which are the result of volcanic activity, its origin is continental – it split from Africa some 6 million years ago. This spawned a great biological diversity, and Socotra is a botanical paradise. Around 800 plant species can be found here – over a third of which are endemic.

The quintessential plant is *Dracaena cinnabari*, the dragon's blood tree. A succulent, it looks like an upturned umbrella and great forests of this strange species can be seen on the rocky slopes of the island's mountainous interior. It is said that Romans travelled to Socotra to

'Bottle tree' at Erher

acquire the tree's blood-red sap, which they used as a disinfectant when they treated wounded gladiators.

The strangely shaped bottle tree – its trunk resembles the distended leg of an elephant – is also found all over the island. It seems to grow everywhere and clusters can often be seen clinging to steep cliffs like watching sentinels. It is also known as the desert rose, as after the monsoon rains it produces shocking-pink flowers.

There are also forests of frankincense trees. Their burnt sap releases a strong fragrance, and was used in ancient Egypt to treat the bodies of the dead. Inscriptions describe how terraces of the trees were cultivated here to meet the demands of this lucrative trade.

The harsh, often dry conditions on the island don't support many animals, but vultures, especially white Egyptian vultures with their bright yellow beaks, seem to be everywhere. They congregate wherever people are eating, to scavenge leftovers, or squat on the low stone houses waiting for their next meal.

The rugged mountains in the centre of Socotra are often enveloped in a misty shroud, which gives the highlands a completely different feel to that of the arid plain encircling them. They are green even during the dry summer months, so this is the time to visit the island if you want to see the greatest contrasts of desert and lush vegetation.

The coast is virtually unspoilt, with long stretches of white sand. At Erher on the east of the island great dunes have formed against high craggy cliffs. The beach extends for kilometres, broken only by fantastically eroded rocks. The shore fringes a brilliant aquamarine

Egyptian vulture

Coastline at Qalansia

Lizard walking over dead coral

Mountainous interior of the island

Frankincense trees in the interior

Dragon's blood trees

lagoon and I saw a large pod of dolphins mooching quietly along, just a metre or so away. Nearby, and set into one of the cliffs, is the immense Huq Cave; more than 3 km long, it is lined with ornate stalactites and stalagmites. Unlike many caves, it is open and airy and it is possible to walk almost its entire length without stooping.

Tourism is very much in its infancy here. The airport was built only in 1999 and apart from a couple of guest houses in Hadibo, the main town, it is a case of camping – or enjoying the hospitality of the local people. And they are certainly hospitable. Many of Socotra's 40,000 or so inhabitants live in isolated villages in the deep fertile valleys of its central mountains, and as you travel around the island you will no doubt often be invited to share their meals. Although most

Rugged valley in the interior of Socotra

Old town of Sana'a, the capital of Yemen

people speak Socotri, the local Semetic language, Arabic is also widely spoken.

Socotra is part of the Republic of Yemen, and if you fly to the island you will have to spend a night in Sana'a, the Yemeni capital. Don't miss the chance to visit its old town, home to what are reputed to be the world's first skyscrapers. These 800-year-old brick buildings soar up to nine storeys and some even incorporate hotels. All the men walk around with traditional curved Jambiya daggers – and often mobile phones – attached to their belts.

Mud-brick building in old Sana'a

ⓘ ···

Yemenia, Yemen's national airline, flies to Sana'a and on to Socotra from a number of countries. Independent travel is all but impossible on the island, and the Yemen-based Universal Touring Company can organize an itinerary for you. November to February are the best months to visit. The strong winds of the south-west monsoon make travel on the island impossible from June to the beginning of September. Supplies of food and bottled water are low in September as supply boats can't run to Socotra from the mainland. Ramadan is also a difficult time to travel.

Tokyo's Parks
Tokyo, Japan

Overlooking the Hama-Rikyu Detached Garden

Although more westernized than Hollywood might have us believe, Tokyo is still a shock of garish neon, clanking _pachinko_ parlours, cutting-edge electronics and incongruously courteous bustle. That much is known the world over. What is less known is that dotted around the city are expansive parks that form tranquil islands of serenity amid Tokyo's seemingly endless development.

In these you will find the lakes and meandering traditional bridges that are important features in Japanese gardens, often with coy carp gliding serenely through dark waters. Herons and cormorants can sometimes be seen fishing or sunning themselves, and turtles occasionally bask on rocks in the middle of the lakes.

If you go to the top of the ultra-modern twin towers of the Tokyo Metropolitan Government Building you will see the city spread out before you – and be able to appreciate the range of the parks, seemingly sprouting among concrete buildings that lap against their edges like waves on the shore. In a city fabled for its modernity, many

Commuters at a pedestrian crossing

Old man painting in Shinjuku Gyoen

of them date back hundreds of years, and some were once reserved solely for the imperial dynasties.

No park shows this better than Hama-Rikyu Onshi Teien. A former royal hunting preserve, and tidally connected to Tokyo Bay, it dates from the 17th century and is ringed by modern skyscrapers. A traditional teahouse sits on a lake in the middle of the park, amid the contemporary reflections. Here a kimonoed waitress will serve you special green tea as you sit at a low table.

This contrast of old and new can also be seen at the Meiji Shrine. Followers of the Shinto religion come from all over Tokyo, often in traditional dress, to pray here, and leave behind wooden *ema* (plaques with wishes inscribed on them). Outside, on the bridge in front of the quiet Inner Garden, the trendy teens of Tokyo gather at weekends, dressed to shock in gothic- and punk-inspired costumes. Although the effect is to make them look rebellious, they pose politely for the ubiquitous photographers.

One of the most centrally located of the parks is the sprawling

Shinjuku Gyoen. It is so large that you can imagine you are in the countryside – not in one of the world's largest metropolises. Birdsong replaces traffic hum, and trees, not buildings, are on the horizon.

Constructed a shade over a hundred years ago, it contains many non-indigenous plants as well as gardens in the French and English classical styles. However, it remains very Japanese. Lawns and plants are clipped to perfection and there are a number of traditional pavilions and bridges. Topiary is big here, and there is even an extensive hothouse where tropical plants flourish under a large glass roof. During the chrysanthemum season the beds are a riot of colour with the flowers perfectly displayed – the blooms are held upright by a network of tiny metal frames. Nature once more bending to the order of Japanese society.

Relaxing in Shinjuku Gyoen

Overlooking the Shinjuku Central Park from the Tokyo Metropolitan Government Building

Traditional tea shop amid modern reflections in Hama-Rikyu Detached Garden

Traditional bridge at the Hama-Rikyu Detached Garden

Shinjuku Gyoen is popular with camera clubs and at weekends it is full of photographers with their tripods capturing every hue.

There are a number of good times to visit Tokyo's parks. Spring, when the flowers begin to bloom, is obviously beautiful, as is autumn, when the leaves start to turn gold and red. The chrysanthemums flower at the beginning of November. Japan becomes collectively hysterical when the cherry trees blossom – their progress is even followed in the news – but the precise time is difficult to predict, although it is generally around the beginning of April.

All Nippon Airways (ANA) flies to Tokyo and also has an extensive domestic route. A good place to stay is the Hotel New Otani, which is situated in its own 400-year-old gardens within the outer moat of the Edo Castle. If you can't afford to stay there the gardens are well worth a visit. For more information about the parks and the best times for various plants and flowers, go to the Tokyo Metropolitan Park Association website.

Cormorant drying its wings in the Hama-Rikyu Detached Garde

Topiary in the Hama-Rikyu Detached Garden

Rapa Nui
Chile

Moai at Ahu Tahai just outside of Hanga Roa

Rapa Nui is one of the most remote inhabited islands in the world. The flight to reach it deep in the south-east Pacific takes you over a seemingly endless expanse of sea, and when you arrive you will feel you have reached the very end of the earth.

The island (also known as Easter Island) is famous for its iconic *moai*, enigmatic stone figures that, over a period of about six hundred years, from the 10th century, were erected on stone platforms, or *ahu*, on various sites along the coast. Many of these locations are now rubble – all that remains of *moai* that were tumbled by malevolent hands as Rapa Nui tore itself apart in internecine conflict, or that have broken up

with the long passage of time. The origins and purpose of these great monuments is unknown, as is the mechanics of how they were transported.

The original inhabitants of Rapa Nui were Polynesians who, it is believed, sailed the Pacific on the trade winds that blew towards the island for half the year, then reversed, allowing these early explorers to find their way home. According to archaeologists, the first settlers came to Rapa Nui to escape the wars, and even cannibalism, that swept the islands of the south Pacific about 1500 years ago. A number of clans were established and the golden age of the creation of the *moai* was ushered in.

A culture thrived in the isolation of Rapa Nui and the population increased to an unsupportable level, peaking in the late 16th century. By

Rano Kao

Rugged coastline on Rapa Nui

this time, however, the islanders had cut down most of the trees, and it is thought that there wasn't enough wood for them to build the boats they needed to leave Rapa Nui. It is also believed that there was famine leading to violence and warfare. The population dropped from a peak of ten thousand to a few thousand over the following 200 years.

There must have been some degree of cooperation in the island's history because the stone for all the *moai* was mined and carved in the same place, before the statues were transported to their platforms, often kilometres away. This place was the spectacular volcanic crater of Rano Raraku, one of two on the island. Here many of the statues can be seen *in situ*. Scores dot its inside and outside, lying at eclectic angles like broken teeth, carved yet awaiting transportation. Others are part-carved in the rock – still waiting for workers to come and finish them, as they have for many hundreds of years. The red stone for the topknots on most of the *moai* was mined from a separate quarry.

Rano Kao, on the north of the island, is the larger of the two craters. Almost a perfect circle, its walls drop vertically for over 100 metres to a lake, speckled with floating reeds, that is rumoured to be twice as deep as the unflooded part of the volcano. On the crater's edge a series of low stone huts provided shelter during the *tangata*

Moai at Anakena

manu, the annual birdman festival. Each clan on the island trained an athlete who competed in a race to climb down the volcano's steep cliffs to the sea, swim to the nearby island of Motu Nui and return with an egg from the *manutara* (sooty tern) in his hand. Ornate petroglyphs record this tradition, which some people believe was a way of distributing scarce resources after the age of the *moai*.

The *moai* are seemingly all over Rapa Nui, and even in the most remote locations. The easiest to reach are on the outskirts of the

Moai at Rano Raraku quarry

Double rainbow over Rano Raraku

small but thriving town of Hanga Roa with its colourful harbour, but probably the most spectacular are those of Ahu Tongariki, lying at the end of a shallow valley that leads down from the Rano Raraku crater. Fully restored, 15 of these great statues stand on a platform; with their crushed noses and cauliflower ears they look for all the world like a lined-up rugby team. And like all but one of the *moai* installations, they are on the coast but staring inland – perhaps prophesying danger from within rather than from the sea.

The landscape of Rapa Nui would be spectacular even without

the *moai*. Apart from the Rano Raraku and Rano Kao craters, a number of volcanic cones are dotted around the island's rolling grasslands, where hundreds of semi-wild horses wander and graze. On the coast, waves that have built up over many kilometres of open sea dash mercilessly against rugged black shores.

On Rapa Nui you can sit in baking sunshine and see a storm in another part of the island. And from a high vantage point you can observe weather fronts sweeping in from the sea. Truly, as you watch the

Wild horse by fallen *moai*, Ahu Tongariki

Petroglyphs at Orongo on Rano Kao volcano

slate-grey clouds dragging the sun in their wake, often with rainbows ahead of them, you will feel that you are on the very edge of the world.

ⓘ ··

LAN Chile is the only airline that flies to Rapa Nui. In peak periods there are daily flights from Santiago and twice weekly from Tahiti. Fares are excessive but cheaper internet deals are available. The small Chilean-run Explora group's package is recommended as it includes all food and drinks, and guided tours around the island with committed and friendly local guides. Rather than relying on motor coaches, the Explora ethos is to conduct a series of walks around Rapa Nui that take in the landscapes and show the *moai* in the best possible context.

Standing *moai* at Ahu Tongariki

1 THE SOCIETY ISLANDS, FRENCH POLYNESIA
2 SI PHAN DON, LAOS
3 NEWFOUNDLAND, CANADA
4 ÎLE DE LA CITÉ, FRANCE
5 ISLANDS OF LAKE TANA, ETHIOPIA
6 PICO RUIVO, MADEIRA, PORTUGAL
7 SAGAR, INDIA
8 VANUATU
9 ISLE OF SKYE, SCOTLAND
10 COUSINE AND PRASLIN, THE SEYCHELLES
11 SANTA BARBARA ISLANDS, USA
12 HONG KONG ISLAND, CHINA
13 AMORGOS, GREECE
14 THE GOLDEN TEMPLE, AMRITSAR, INDIA
15 THE FLORIDA KEYS, USA
16 BALI, INDONESIA
17 SOUTHERN DALMATIAN ISLANDS, CROATIA
18 YASAWA ISLANDS, FIJI
19 MADAGASCAR
20 STOCKHOLM, SWEDEN

21 LORD HOWE ISLAND, AUSTRALIA
22 SRI LANKA
23 SARK, UNITED KINGDOM
24 BACUIT ARCHIPELAGO, PALAWAN
25 LAMU, KENYA
26 BIG ISLAND, HAWAII, USA
27 ST LUCIA, WINDWARD ISLANDS
28 MONT SAINT-MICHEL, FRANCE
29 TIERRA DEL FUEGO, ARGENTINA
30 SICILY, ITALY
31 MOUNT DESERT ISLAND, MAINE, USA
32 SVALBARD, NORWAY
33 ELEUTHERA, THE BAHAMAS
34 IBIZA AND FORMENTERA, SPAIN
35 KUBU ISLAND, BOTSWANA
36 ISLES OF SCILLY, ENGLAND
37 PHANG NGA BAY ISLANDS, THAILAND
38 SOCOTRA, YEMEN
39 TOKYO'S PARKS, JAPAN
40 RAPA NUI, CHILE

French Polynesia
Tahiti Tourisme
www.tahiti-tourisme.com
Air Tahiti Nui
www.airtahitinui.com
Mahana Dive
www.mahanadive.com

Thailand and Laos
Tourism Authority of Thailand
www.tourismthailand.org
Etihad Airways
www.etihadairways.com
Audley Travel
www.audleytravel.com
Paradise Koh Yao
www.theparadise.biz

Newfoundland
Newfoundland and Labrador
Information
www.nfld.com

Paris and Mont Saint-Michel
Paris-Île-de-France Tourist Board
www.pidf.com
Eurostar
www.eurostar.com
Aurigny Air
www.aurigny.com

Ethiopia
Ethiopian Airways
www.flyethiopian.com
Journeys By Design
www.journeysbydesign.co.uk

Madeira
Madeira Tourism Board
www.madeiratourism.org
Reid's Palace Hotel
www.orient-express.com

India
India Tourism
www.incredibleindia.org
Atithi Voyages
www.atithivoyages.com
Jet Airways
www.jetairways.com

Vanuatu
Vanuatu Tourism
www.vanuatutourism.com
Air Vanuatu
www.airvanuatu.com
Destination Pacific
www.destinationpacific.vu
Le Meridien, Port Vila
www.starwoodhotels.com/
lemeridien

Scotland
Bosville Hotel
www.bosvillehotel.co.uk

Cousine and Praslin
Cousine Island
www.cousineisland.com
Turquoise Holidays
www.turquoiseholidays.co.uk
Lemuria Resort, Praslin
www.lemuriaresort.com
Air Seychelles
www.airseychelles.com

Santa Barbara Islands
Catalina Island Chamber of
Commerce and Visitors Bureau
www.catalinachamber.com
Island Packers
www.islandpackers.com
Catalina Island Inn
www.catalinaislandinn.com
Catalina Express
www.catalinaexpress.com

Hong Kong
Hong Kong Tourism Board
www.discoverhongkong.com
Mandarin Oriental Hotel
www.mandarinoriental.com

Amorgos
Aegialis Hotel
www.amorgos-aegialis.com

Florida Keys
Florida Keys and Key West
www.fla-keys.com

Bali
Aman Resorts
www.amanresorts.com

Croatia
Sail Croatia
www.sailcroatia.net

Fiji
Fiji Visitors Bureau
www.bulafiji.com
Nanuya Island Resort
www.nanuyafiji.com

Madagascar
Le Voyageur
www.madagascar-tour-
operator.com
Air Madagascar
www.airmadagascar.com

Stockholm
Visit Sweden
www.visitsweden.com
Stockholm Visitors Board
www.stockholmtown.com
SAS
www.flysas.com
Nordic Light Hotel
www.nordiclighthotel.se
The Archipelago Foundation
www.skargardsstiftelsen.se
Waxholmsbolaget
www.waxholmsbolaget.se

Lord Howe Island
Lord Howe Island Tourism
Association
www.lordhoweisland.info
Capella Lodge
www.lordhowe.com

Sri Lanka
Sri Lanka Tourism
www.srilankatourism.org
Jetwing Hotels and Travels
www.jetwingtravels.com
Emirates
www.emirates.com

Sark
Sark Tourism
www.sark.info
Aurigny
www.aurigny.com

Palawan
Philippines Department of Tourism
www.wowphilippines.co.uk
El Nido Resorts
www.elnidoresorts.com
Manila Pavilion Hotel
www.manilapavilion.com.ph

Kenya
Air Kenya
www.airkenya.biz
Peponi Hotel, Lamu
www.peponi-lamu.com
Ethiopian Airways
www.flyethiopian.com

Hawaii
Hawaii Visitors and Convention
Bureau
www.gohawaii.com
Mauna Lani Resort at Kalahuipua'a,
Big Island
www.maunalani.com
Outrigger Keauhou Beach Resort,
Kailua-Kona, Big Island
www.outrigger.com
Hilo Hawaiian Hotel, Hilo, Oahu
www.castleresorts.com
Ala Moana Hotel, Honolulu, Oahu
www.alamoanahotel.com

St Lucia
St Lucia Tourist Board
www.stlucia.org
Coco Palm Hotel
www.coco-resorts.com

Tierra del Fuego
Cruceros Australis
www.australis.com
South American Experience
www.southamericanexperience.co.uk

Sicily
Tourist Information
www.bestofsicily.com

Mount Desert Island
Balance Rock Inn
www.balancerockinn.com
Colgan Air Inc.
www.colganair.com
Atlantic Oakes Hotel
www.barharbor.com

Svalbard
GAP Adventures
www.gapadventures.com
SAS
www.flysas.com

The Bahamas
Bahamas Tourist Office
www.bahamas.com
Virgin Atlantic Airlines
www.virgin-atlantic.com
Bahamasair
www.bahamasair.com
Coral Sands Hotel
www.coralsands.com
Radisson Cable Beach and Golf
Resort
www.cablebeachresorts.com
Bahamas Fast Ferry
www.bahamasferries.com

Ibiza and Formentera
Fomento del turismo de Ibiza
www.illesbalears.es
Agroturisme Can Lluc
www.canlluc.com

Botswana
Uncharted Africa
www.unchartedafrica.com

Isles of Scilly
Isles of Scilly Tourist Board
www.simplyscilly.co.uk
Island Hotel, Tresco and Hell Bay,
Bryher
www.tresco.co.uk
British International Helicopters
www.islesofscillyhelicopter.com
Star Castle Hotel, St Mary's
www.star-castle.co.uk

Yemen
Universal Touring Company
www.utcyemen.com
Yemenia Yemen Airways
www.yemenia.com

Tokyo
Japan National Tourist Organization
www.jnto.go.jp
ANA
www.anaskyweb.com
Tokyo Metropolitan Park
Association
www.tokyo-park.or.jp/english

Rapa Nui
Casa Rapa Nui
www.explora.com

1 3 5 7 9 10 8 6 4 2

Published in 2007 by BBC Books, an imprint of Ebury Publishing.
Ebury Publishing is a division of the Random House Group Limited.

Text copyright © Steve Davey 2007

Steve Davey has asserted his right to be identified as
the author of this Work in accordance with the Copyright,
Designs and Patents Act 1988.

Photographs copyright © Steve Davey and Marc Schlossman 2007

The Random House Group Limited Reg. No. 954009
Addresses for companies within the Random House Group
can be found at www.randomhouse.co.uk

A CIP catalogue record for this book is available from the
British Library.

ISBN 978 0 563 49351 8

The Random House Group Limited makes every effort to ensure that
the papers used in our books are made from trees that have been
legally sourced from well-managed and credibly certified forests. Our
paper procurement policy can be found at www.randomhouse.co.uk

Commissioning editor: Nicky Ross
Project editor: Christopher Tinker
Copy-editor: Tessa Clark
Designer: Bobby Birchall at Bobby & Co, London
Production controller: Kenneth McKay

Set in DIN Regular
Colour origination and printing by
Butler & Tanner Ltd, Frome, England

www.stevedavey.com
www.marcschlossman.com

This book is dedicated to my daughter, who Katharine and I somehow managed to conceive in the middle of such a hectic shooting schedule. *SD*

Acknowledgements

A project such as this is a collaborative effort, and relies on the help and generosity of many people and organizations from all over the world.

Steve would like to thank, in order of shooting, and with apologies to anyone inadvertently omitted: Paris-Île-de-France Tourist Board; Eurostar; Madeira Tourism Board; Orient Express Hotels; Hong Kong Tourism Board; Mandarin Oriental Hotel, Hong Kong; Japan National Tourist Organization; ANA; Cousine Island; Turquoise Holidays; Lemuria Resort, Praslin; Air Seychelles; Air Kenya; Peponi, Lamu; Ethiopian Airways; Journeys By Design; Indiatourism, London; Atithi Voyages; Jet Airways; District Magistrate Roshni Sen; Golden Temple, Amritsar; Tourist Authority Thailand; Etihad; Audley Travel; Paradise Koh Yao; Air Tahiti Nui; Tahiti Tourism; Mahana Dive, Huahine; Casa Rapa Nui, Explora Hotels; Air Vanuatu; Vanuatu Tourism; Destination Pacific; Le Meridien, Port Vila; Fiji Visitors Bureau; Nanuya Island Lodge, Fiji; Capella Lodge, Lord Howe; NSW Tourism; Aman Resorts, Bali; Uncharted Africa; Le Voyageur, Madagascar; Air Madagascar; GAP Adventures; the crew of the MS *Explorer*; Scandinavian Airlines; Sail Croatia; Aurigny Air; Sark Tourism; Tresco Estates, Isles of Scilly; Star Castle Hotel, St Mary's; Scilly Tourism; British International Helicopters; Sri Lanka Tourism Office; Jetwing Holidays; Fomento del turismo de Ibiza; Agroturisme Can Lluc; Ibiza; UTC Yemen; Yemenia Yemen Airways and the following PR companies who helped organize many of the trips: Lush PR; Indigo PR; Mango PR; Hills Balfour; Saltmarsh PR; Seal Communications; BGB Associates; PR Co; Oliver Relations; Southern Skies and Cut Communications. All the pictures in this book were shot on the Nikon D2x. Thanks to Jacobs, New Oxford Street for selling them to Steve and to Nikon UK Professional Service for mending them when he breaks them.

Steve would also like to thank his family and friends for still being there when he got back from his many trips, and Katharine, who had to go through the first six months of pregnancy virtually on her own while he was on the road. He would also like to thank Marc Schlossman, the associate photographer for this book, for his hard work and professionalism.

Marc would like to thank Cruceros Australis; South American Experience; Coral Sands Hotel, Harbour Island; Radisson Cable Beach and Golf Resort, Nassau; Virgin Atlantic Airlines; Bahamasair; Bahamas Tourist Office; J. J. Angove Ltd; Aegialis Hotel, Amorgos; El Nido Resorts, Palawan; Philippines Department of Tourism; Coco Palm Hotel, St Lucia; St Lucia Tourist Board; Bosville Hotel, Skye; Nordic Light Hotel, Stockholm; Scandinavian Airlines; The Archipelago Foundation; Östanviksgard, Nämdö; Stockholm Visitors Board; Visit Sweden; Island Packers, Ventura; Catalina Island Inn, Avalon; Catalina Island Chamber of Commerce and Visitors Bureau; Mauna Lani Resort, Big Island; Outrigger Keauhou Beach Resort, Big Island; Hawaii Visitors and Convention Bureau; Balance Rock Inn, Bar Harbor; Colgan Air Inc.; Atlantic Oakes Hotel, Bar Harbor; Michael Patrick Destinations and Communications Ltd.

Marc would also like to thank Nici, Ben and Theo for love, patience and being there over many long journeys. And to Steve: thank you. I am proud to be your colleague and friend.

Both Steve and Marc would like to thank everyone at BBC Books for their vision and professionalism, notably Nicky Ross, Christopher Tinker, Stuart Biles and Kenneth McKay, and also Tessa Clark, Bobby Birchall and Vicki Vrint. They would also like to thank all of the guides, who showed them their countries with such enthusiasm and generosity and all the individuals who contributed, but unfortunately there isn't enough space to name personally.

Marc Schlossman shot Palawan, Santa Barbara, Tierra del Fuego, Florida Keys, St Lucia, Hawaii, Stockholm, Skye, Sicily, Amorgos, Mount Desert Island, the Bahamas and Newfoundland. Steve and Marc shot Sark and Mont Saint-Michel together.